What People Are Saying About

How to Run a Planet

John Holman has written an extraordinary book that should be read and debated by anyone concerned with the world as it is, and who believes in the necessity, possibility, and desirability of adopting radically better ways of organizing and sustaining our collective, planetary existence. The reliance on esoteric thought gives Holman's worldview a challenging originality that exposes the shortcomings of rationalism and materialism.
Richard Falk, Professor Emeritus of International Law, Princeton University, and Chair of Nuclear Age Peace Foundation

A work that should be examined by all interested in building a peaceful world.
Douglas Roche, Former Canadian Senator and Chair of United Nations Disarmament Commission

A lovely and important book that asks us to reject the main narrative of economic globalization, unrestricted markets and unlimited growth — followed by most governments — and to embrace instead one based on inclusivity, sustainability, spiritual growth and a governance system geared to achieving social and environmental well-being for all.
Maude Barlow, Co-Founder of the Council of Canadians and Chair of Food & Water Watch

How to Run a Planet

An Essential Guide to the Big Picture and Global Solutions

Previous Books

The Return of the Perennial Philosophy: The Supreme Vision of Western Esotericism

ISBN: 9781905857586

How to Run a Planet

An Essential Guide to the Big Picture and Global Solutions

John Holman

CHANGEMAKERS BOOKS

London, UK
Washington, DC, USA

First published by Changemakers Books, 2025
Changemakers Books is an imprint of Collective Ink Ltd.,
Unit 11, Shepperton House, 89 Shepperton Road, London, N1 3DF
office@collectiveinkbooks.com
www.collectiveinkbooks.com
www.changemakers-books.com

For distributor details and how to order please visit the 'Ordering' section on our website.

ISBN: 978 1 80341 704 2
978 1 80341 715 8 (ebook)
Library of Congress Control Number: 2023949854

A CIP catalogue record for this book is available from the British Library.

Design: Lapiz Digital Services

UK: Printed and bound by CPI Group (UK) Ltd, Croydon, CR0 4YY
Printed in North America by CPI GPS partners

Contents

Introduction ix

Part I: The Big Picture 1

Chapter 1: Our Place in the Universe 3

Introduction 3
Living Sentient Entities 4
Horizontal and Vertical Universes 5
Higher Kingdoms of Nature 6
Authors and Characters 7
Who Am I? 8
The Holarchical Universe 11
What Am I Here For? 12

Chapter 2: The Nature of Things 14

Introduction 14
Spheres within Spheres 15
The Consciousness Cycle 16
Esotericism, Metaphysics, and Science 17

Chapter 3: Past-Present-Future 24

Introduction 24
Aevertinity 25
Consciousness-Time 25
World Periods 29
How Am I to View the Universe? 33
Coda 36

Part II: Global Solutions 39

Chapter 4: Good Global Government 41

Introduction 41
The Purpose of Government 42
The Priorities of Government 44

Excerpts from the Universal Declaration of Human Rights 46
The Problems with Existing Government 48
The Global Democracy Manifesto 53
Spiritual Aristocracy 56
World Government 58
Holarchy and Subsidiarity 65
Chapter 5: Economics, History, Society **68**
Introduction 69
The Prenatal and Postmortal Ages 70
The Infant and Pensioner Ages 72
The Child Age 75
The Youth Age 77
The Young Adult Age 78
The Mature Adult Age 81
The Old Adult Age 89
Chapter 6: The Way Forward **93**
Introduction 93
Wellbeing, Not Economic Growth 99
Beyond the 2030 Agenda 103
Five Common but Mistaken Economic Views 105

Notes 111

Introduction

> Billions of our citizens continue to live in poverty and are denied a life of dignity. There are rising inequalities within and among countries. There are enormous disparities of opportunity, wealth and power. Gender inequality remains a key challenge. Unemployment, particularly youth unemployment, is a major concern. Global health threats, more frequent and intense natural disasters, spiralling conflict, violent extremism, terrorism and related humanitarian crises and forced displacement of people threaten to reverse much of the development progress made in recent decades. Natural resource depletion and adverse impacts of environmental degradation, including desertification, drought, land degradation, freshwater scarcity and loss of biodiversity, add to and exacerbate the list of challenges which humanity faces. Climate change is one of the greatest challenges of our time and its adverse impacts undermine the ability of all countries to achieve sustainable development. Increases in global temperature, sea level rise, ocean acidification and other climate change impacts are seriously affecting coastal areas and low-lying coastal countries, including many least developed countries and small-island developing States. The survival of many societies, and of the biological support systems of the planet, is at risk.

The above is taken from the United Nations 2030 Agenda for Sustainable Development.[1] It was written in 2015, and since then increasing extreme weather events, the Covid-19 pandemic, the wars in Ethiopia, Ukraine, and Israel-Palestine, and the rise or resurgence of populism and ultra-nationalism have only highlighted our environmental and political crises. It is in response to these crises that this book is written. It is entitled *How to Run a Planet* in the sense that it offers a guide to creating a better world, but it may also be taken as a treatise on society and nature. Why the latter too? Because what doesn't help us in our political task is a historically recent, but now globally dominant, *materialist* worldview. A worldview which says that the universe is purposeless and material, and that politics is a mere temporal concern.

This book rejects the materialist worldview and invites all sensible people to do the same. It also rejects capitalism and pessimism. It says materialism is misguided, capitalism is immature, and pessimism — while understandable when we look at the daily news — is ultimately unwarranted. Taking the longest view of human history, we can see a slow ascent of civilization toward a world of peace, unity, and wellbeing. The book could be seen to be in the tradition of a recurring philosophy that answers the question "How are we to run our planet?" with the short reply "In a way consistent with the cosmic order." By "cosmic order" is meant the way the universe is and operates. But here's the thing: that cosmic order, so this philosophy says and this book advances, is best represented in the cosmology of esotericism, not in the lesser big picture presented by modern, conventional Western science.

The enlightened skeptic will ask at this point: "What is esotericism and why should I entertain it?" In terms of the "what," esotericism is the tradition of thought that predates classical antiquity and includes in its family of members the schools of Pythagoreanism, Platonism and Neoplatonism, Hermeticism, Gnosticism, Kabbalism, Theosophy, and Traditionalism. As a recurring presence it has earned the designation "Perennial Philosophy." Like any family, the members of this tradition sometimes disagree over particular matters, but there is general agreement on a picture of the universe featuring God (or an equivalent generating principle) and a specific and noble task for humanity.

The word "esoteric" means inner or hidden. Esotericism as a tradition of thought is the philosophical and mystical tradition distinguishable in the West (without being exclusive to it) that speaks of inner or hidden truths regarding the universe. Within this tradition is another meaning of esotericism — as the contemplative practice that leads to discovering these truths. Both esotericisms are discussed in some detail in the author's previous book.[2] It might be said that when it comes to the "science versus religion" debate, particularly in the form of the new atheism versus the Abrahamic religions, esotericism is the older and enigmatic third voice. A voice that is not normally invited to the debate as it is accommodating to both science and religion — just not to either scient*ism* or religion*ism*. *The Cambridge Handbook of Western Mysticism and Esotericism* defines esotericism as:

> The hidden intellectual history of the West, running like a dark thread through the fabric of the more conventional intellectual history we have all been taught. The influence of mystics and esotericists on science, philosophy, theology, literature, politics, and popular culture is immense, but it is a story scholars are only just beginning to tell ... 'Esotericism' refers to a number of theories, practices, and approaches to knowledge united by their participation in a premodern, largely pagan worldview. Central to this worldview is commitment to the idea of the unity of existence — that existence is an interrelated whole in which seemingly dissimilar things exist in qualitative correspondence and vibrant, living sympathy. The ruling correspondence is 'as above, so below': The objects that surround us (and their relationships)

> mirror, in a fashion that can be called 'emblematic', the fundamental features of the universe as a whole. Most important of all, we mirror those features in our own bodies and souls. These correspondences are discovered through the cultivation of supernormal aspects of human subjectivity ... Esotericists typically hold that such knowledge can be utilized to effect changes in the world or in the self through causal mechanisms that empiricism finds inexplicable (and, therefore, rejects as impossible). This commitment usually goes hand in hand with the belief that the same supernormal aspects of the subject can reveal the existence of other dimensions of reality, usually hidden from view. Further, esotericists typically believe that the truths and practices just mentioned are of the greatest antiquity — perhaps once widely disseminated and openly proclaimed, but now (and for a great many centuries) hidden and preserved by a few special individuals or schools. Discovery in esotericism is almost always rediscovery.[3]

The phrase "as above, so below" is almost the catchphrase of esotericism and particularly relates to (or can be applied to) the work of government, as will be drawn out in Part II of this book. Special note should be made of the phrase "cultivation of supernormal aspects of human subjectivity." This is central to esotericism and distinguishes it as the "path of gnosis" (where *gnosis* refers to direct knowledge of the inner or hidden truths through contemplation), as distinct from the "path of reason" (science — or knowledge through ratiocination and experiment) and the "path of faith" (religion — or knowledge through faith).

The renewed academic and popular interest in esotericism in recent decades (the latter in the form of New Age religion and Neopaganism) coincides with the crisis of the West, referring to the growing rejection of mainstream Western thought — at the very moment of its global "triumph" — in the form of its economic and scientific "wisdom." The former includes the endless pursuit of economic growth based on the exploitation and degradation of the natural world and the human being alike. The latter includes a physical cosmology in which existence is, we are told, purposeless, and the universe is made up of matter and energy, most of it invisible, out of which somehow sprang life and consciousness. Esotericism's cosmology is a vaster and more satisfying big picture — more in keeping with the "Grand Design" in Eastern (and particularly Hinduist) thought, and in the teachings of many indigenous peoples, as Table 1 introduces.

Table 1: The Big Picture

Question	Esotericism	Conventional Western science
Is there a God?	Yes, or an equivalent generating principle	No, or need not be
What is the universe made of?	Consciousness/life	Matter/energy
Is there a purpose to existence?	Yes, to express the Good	No
When did the universe begin?	In the present	In the past
Is the universe ordered in a way that provides a model for good global government?	Yes — as above, so below	No, or "pass"

Esotericism rejects materialism in favor of idealism, and both evolutionism and creationism in favor of emanationism (as defined in Chapter 1). It also rejects the Big Bang theory, and the idea of a finite end to a purposeless universe, in favor of a cyclically manifesting universe with no beginning and no end but with purpose and with progression. Esotericism rejects atheism and agnosticism in favor of panentheism and gnosticism, and it rejects a narrow naturalism and mechanism in favor of a broader naturalism and organicism. Some political movements left and right have invoked esotericism for their own ideological ends, but the tradition itself takes no position on the modern political spectrum because this is rooted in the materialist worldview which esotericism rejects.

Part I of this book presents the cosmology of esotericism in a picture that should be familiar to scholars and practitioners of esotericism alike. It is fair to say that the author is both and that he is persuaded this cosmology is valid in the sense of being contemplatively well-grounded. It is, of course, for the reader to make up his or her own mind about this. Acceptance of this cosmology is not a requirement for entertaining the political ideas in Part II, which are congruent with the worldview of esotericism without being exclusive to it. The aspiritual reader is free to skip to Chapters 4 and 6 — the "how to" parts per se — if s/he wishes, although as this would miss the soul of the book, the key "as above, so below" principle, and only an open mind is required, that would be regrettable.

Which takes us to the "why" question. It is customary perhaps to think cosmology is irrelevant when it comes to politics, but as the cultural historian Theodore Roszac reminded us, "Cosmology implicates values." A mindless and heartless picture of the universe (such as the materialist picture is) obviously works against having a mindful and compassionate society. There is, indeed, a direct correspondence between a modern (post-Enlightenment) natural philosophy which says that the universe is material and operates by its own physical laws only, with no need for a spirit behind and directing it toward an end other than spatial expansion, and a modern political philosophy which says that society is animalistic and operates by its own economic laws only, with no need for a government behind and directing it toward an end other than wealth expansion (economic growth). Current economic globalization is not working for ordinary people and the planet, not because we haven't sufficiently liberalized the economy yet (as those on the Right say), but because there is not yet a government behind and directing the economy for the good of ordinary people and the planet. Getting economic globalization right depends on getting political globalization right, and both depend on getting cultural globalization right — which in the end refers to the need for a common *non*materialist worldview. This concern with materialism was keenly identified by the futurist Willis Harman:

> Since modern society ascribes no 'reality' to inner experience, transcendent values have no power and materialistic values prevail. Thus it seems reasonable for society to be characterized by economic rationalization of an ever-increasing fraction of social behaviour and organization. Industrialization of production of goods and services gradually extends to more and more of human activities; increasingly, they all become included in the economy. One result is monetization and commercialization (all things coming to be measurable by and purchasable in units of currency). The economic rationalization of knowledge leads to the "knowledge industry": to science justified by the technology it produces, and to education justified by the jobs it prepares for. Economic rationality becomes predominant in social and political decision-making, even when the decisions it leads to, are unwise by other standards (such as the wellbeing of future generations). Technological solutions are attempted for problems that are basically socio-political in nature. The worth of persons (to say nothing

> of our nonhuman fellow creatures on earth) is assessed by their value in the economy. Humankind's relationship to the earth is essentially an exploitative one.[4]

And addressing the area of global governance in particular, the Executive Director of the Global Governance Forum, Augusto Lopez-Claros, writes:

> The most fundamental area in which to build consensus is the ultimate purpose of governance itself. Why do we need governance: to avoid anarchy and collective destruction, and/or to achieve some common purpose? ... Those with a purely materialist orientation would say it is sufficient for everyone to be properly fed, well housed, comfortable, in some security and appropriately distracted to achieve some superficial form of happiness. However, most cultures, all religions and many psychologists accept an ethical framework and a concept of a higher fundamental purpose and potential for the human person than just material satisfaction. If our goal is a more just and sustainable social order, we shall need to cultivate other qualities of character such as moderation, justice, love, reason, sacrifice and service to the common good, which are necessary to overcome the focus on ego, greed, apathy and violence behind many of the failures of governance today. At the highest level, effective governance should enable each person to develop her or his inherent potential and to refine her or his character, while contributing to the advancement of civilization. For most of humanity, human experience is ultimately spiritual in nature, and cultivating our higher qualities and the endless potential in human consciousness is at the core of what it is to be human.[5]

There is thus a direct link: a) between our worldview and the achievement (through good global government) of societal and environmental wellbeing, and b) within our worldview, between natural and political philosophy. It takes a village to raise a child, they say. Well, it also takes a society to have the right worldview to solve the problems caused by having the wrong or misguided one. What this book puts forward is a schema that is actually more faithful to a historical global consensus than the Western view that originated in the seventeenth century and which stuck in a lot of minds in the twentieth century. A key part of this proposition is a return to "verticality" in cosmology, mirrored in the architecture of global governance, as will be explained. The holding on to a "horizontality only," both in our worldview and politics, is non-progressive.

The reader will be taken into deep metaphysical waters in Part I. Those who are altogether unfamiliar with esotericism and/or are more interested in the political guide per se may wish to jump straight to Part II as said, although as there is no requirement to take up meditation or embrace a theistic view, the reader will find jumping in refreshing. Plus, we do generally recognize that our desacralized big picture of the universe is complete with knowledge but incomplete with answers, do we not? And that there remains a huge longing for *meaning* which is just not satisfied by physics textbooks, technological advancements, and commercial and consumerist pursuits. This isn't a "the universe is vast and wonderful but ultimately indifferent to us selfish grasping sapiens" type book — quite the opposite.

All in all, this book responds to the responsibility that "value-oriented intellectuals" have (as Noam Chomsky reminds us) for trying to find solutions to the world's problems — problems which are as much philosophical as they are political. The late physicist Stephen Hawking believed that philosophy was dead. Well, the author, as someone influenced by the love of wisdom (as opposed to those analytic or continental philosophies which cede reality to empirical science and in so doing nullify themselves), could not *dis*agree more, asserting instead that as we move further into the twenty-first century it is more a case of "scientism is dead." The practice of science will continue indefinitely, but we cannot say the same of the conventional Western scientific view, which is increasingly being accepted (and not just by cultural historians and postmodern philosophers but also by astrophysicists as they struggle to explain time and the "missing 95%") *as* just a view, that is, one that is not necessarily reflective of reality. This being so, there is no necessary reason why we should continue to accept it as our standard civilization view — we are permitted, if not obligated, to consider other views.

It is estimated that by 2050 only 8% of the world's population will be from the West, and it is reasonable to think that a new *planetary* civilization in the decades or centuries ahead requires a worldview that is truly global, that is, inclusive also of Eastern and indigenous peoples. Esotericism allows a dialogue in our professional and social lives with the world's major faiths (one of which most people do identify with), which share some belief in a "divine plan," including a work of "creating Heaven on Earth" — a job which would involve replicating some celestial model in earthly government, and citizens following the "Golden Rule" (treat others as you would have them treat you), in recognition of a "one family" truth. This interfaces with the "Way of Nature" and the "Great Unity" in Taoist and Confucianist thought, and with *Vasudhaiva Kutumbakam* in Hinduist thought.

Esotericism allows a dialogue too with the environmental movement, proved 100% right and leading the way in policy prescription and action, if perhaps lacking a driver other than ecological necessity for decarbonizing and degrowing. The missing driver is, or links with, wellbeingism as covered in

Chapter 6: doing the right thing not just because we have to, but because we want to. If there is a need (and there is) for going "part-time" as a civilization, then that needs to be positively framed as allowing for the maximization of wellbeing and self-realization; we are talking here about human rather than economic growth, and sustainable living rather than sustainable development which is something of an oxymoron. And finally and perhaps crucially, esotericism allows for a dialogue with young people who are idealistic and socially aware, cosmopolitan and planetarist, but prey to pessimistic and nihilistic gloom. There's no shortage of things to be alarmed and depressed by, and yet the post-World War II international organizations still stand and are defended, global justice and net zero are on the agenda, and there is an indefatigable will to realize a better future. The existence of this will in itself points to an evolution of consciousness (as covered in Chapter 5) beyond a juvenile idolatry of wealth and nation-states.

Esotericism has informed the development of Western thought generally and explains the deeper thought of Pythagoras and Plato. The thought of these two often comes to us diluted through the summaries of contemporary scholars, with a tendency to assess them as the product of an intellectually inferior past. But as with a Stradivarius violin or a Chippendale chair, sometimes the superior is in the past. And in any event, we are not talking about de-evolving reason, or going backwards to medieval religiosity, or giving up the benefits or halting the advance of science and technology. Rather, we are talking about *relativizing* the conventional scientific view and not just culturally-historically: ultimately, appreciating it as linked to only one mode of awareness (or type of consciousness or aspect of human subjectivity – hence our flagging that phrase earlier) – thereby redefining scientific knowledge and repositioning science as a cultural activity. More on this in Chapter 2 as we consider "Type 1" and "Type 2" knowledge.

The fundamental inequity in our world society, manifested between countries and between individuals and communities within countries, is the number one source of political instability and influences global conflicts. Immiseration and frustration inevitably lead to social unrest: it's not rocket science. An analogue is what happens in the human body if the oxygen- and nutrient-containing blood does not reach all parts of the body. As Paul Kennedy, Director of International Security Studies at Yale University, says, we have permitted this situation to arise, meaning there must be something in our worldview which tolerates such inequity:

> This would be intriguing to an extra-terrestrial observer ... it rarely occurs to us to think that all the earth's other species — be it sheep, cod or sparrows — have roughly the same standards of living and consume roughly the same amounts as every other sheep, cod or sparrow each day. But human beings have permitted a situation to arise in which certain of their societies enjoy levels of consumption 200 times greater than other societies.[6]

Deep thinkers who have studied this problem are unanimous. If we see each other as, well, *other* (essentially separate selves, not selves that are coincident in essential identity), then there is no "basis in nature" for ensuring social justice. We may on one level appreciate what makes good sense politically, which is managing wealth and resources distribution to ensure that equity, but on a deeper level we see selfishness and greed as "natural." Here is that unhelpful materialist worldview — the "technocratic paradigm" as Pope Francis calls it[7] — providing an obstacle to change in itself, and making us dismissive of sharing and caring visions which we wrongly see as unrealistic and even unnatural.

The same can be said when it comes to the environment. If we see no purpose to existence — and certainly no purpose to express the Good — and we see the universe as just made up of matter and energy (material for our "naturally" selfish and greedy selves to degrade and exploit), then while we may on one level appreciate what makes good sense ecologically — which is tackling climate change and being proper guardians of nature — at a deeper level our worldview is antipathetic toward, even hostile toward, carrying out this work. This is why the philosophical and the political (the first including metaphysics and the second including economics) cannot be separated.

A key aim of this book is to offer a map and a compass in troubled times. A philosophical map, that would be, covering the cosmological, ontological, and teleological context to our political task. And a compass to guide the actions of policymakers, activists, and leaders over the next few decades. The world cannot be transformed overnight, and things currently in play have to play themselves out, but significant progress can be made by the end of this century. That we don't currently have a map and compass of this kind is axiomatic. Without subscribing to neoliberal economic globalization, which most agree is harmful, the author would assert that globalization generally is a good thing as a step toward a planetary civilization. He would assert also that today's positive transformation movements (for social justice, global democracy, and a more holistic worldview) are the fetal movements of that planetary civilization to come which will be genuinely *supra*modernity and not just *post*modernity.

Postmodernity is the period we are currently in, in which the old man (modernity) looks back over his life and, while recognizing his achievements (scientific and technological progress), realizes he put work before family (society and nature). *Supra*modernity would be a new incarnation of the man — perhaps as a woman. Modernity brought with it, as well as the problems now giving rise to the crisis of the West, the establishment of sovereign nation-states. This would be intriguing also to Kennedy's extraterrestrial observer and is discussed as the subject of world government in Part II – what happens *after* globalization, as this is a journey of integration but not the end destination of that journey. It is not spoiling the plot to say that this book supports the proposal, endorsed by the European Parliament and growing numbers of politicians worldwide, for a United Nations Parliamentary Assembly.

Part I
The Big Picture

Chapter 1

Our Place in the Universe

Introduction

There is a story that when Captain Cook first landed in Hawaii he said to a native, pointing his finger, "What's that over there?" "Pele," replied the native (Pele being the Hawaiian goddess of fire). "You mean a volcano in which your goddess lives?" asked Cook. The native looked at him, puzzled. "No, it's Pele."

The lesson of this story is that here we had two people looking at the same thing, but because they grew up in different cultures they saw different things. Indeed only one of them (Cook) saw a *thing*, a natural feature, a volcano. The other saw an *entity*, a nature spirit, a goddess.

"I believe what I see," people say. But do they/we? Do we see, rather, what we believe; what we were *enculturated* to see? Is that a tree in your garden or is it a nature spirit? If you say "tree," is that merely because you grew up in a culture that sees it as a tree? What about the river over there: is that just a natural flowing watercourse or is it a living being in some way? How about the Earth: is it just a rock we live on, or a god we live in? Is the universe just matter and energy which forms into astronomical objects and bodies, or is it a living — and sentient — entity within which galaxies, stars, and planets are also living sentient entities? What do we *really* know about what exists and its nature beyond what we were taught to believe and the pre-assumptions of scientific materialism? The oldest religion in the world is animism — the belief that everything is alive, conscious in some way, and has a nonmaterial essence which may permit us to call it divine. This is the basis of esoteric thought.

Living Sentient Entities

We find ourselves in existence. In that, we have no choice, but otherwise we have free will, says esotericism, and in the conventional scientific view the Earth has a place in the universe, which is a physical position in space. We are in the Sol solar system, which is in the Orion arm of the Milky Way galaxy, which is in the Laniakea Supercluster of galaxies. In the esoteric view this spatial positioning of the Earth is not wrong exactly, but the conventional appreciation of what these and other such astronomical bodies/objects are, and their relation to one another (and by extension our relation to them), *is* wrong. Or to put it another way, it is a case of not seeing the full picture.

Many ancient philosophers saw planets and stars as "animals" — not literally and specifically as biological organisms, but metaphorically and generally as living sentient entities (we might use the acronym LSE). This is correct, says esotericism, although it adds that we can only partly compare the nature of their conscious existence with that of our own, for it is qualitatively distinct. And a direct, if still limited, appreciation of their sentient nature (as opposed to an indirect, philosophical one) depends on a supernormal mode of awareness. Click on the conventional mode of awareness, so to speak, and the conventional appreciation is had. Click on that supernormal mode of awareness, and the esoteric appreciation is had. Regarding modes of awareness, the Traditionalist philosopher Seyyed Hossein Nasr wrote:

> Science is based in fact upon the idea that there is only one mode of perception and one level of external reality which that single level of consciousness studies. The world according to it is what we see if we extend the word "see" to include what is shown by the microscope and the telescope which do not represent a new mode or level of seeing but simply the extension, horizontally, of what the

human eye perceives. In contrast, authentic spirituality is based upon the basic thesis that not only are there levels of reality but also levels of consciousness that can know those levels of reality. What we perceive of the external world depends upon our mode of consciousness.[1]

Horizontal and Vertical Universes

In the language of the Traditionalist school of esotericism, the conventional appreciation is of a "horizontal" universe. The esoteric appreciation is of a "vertical" universe. These two correspond to the two universes in ancient philosophical thought: 1) the mundane or sensible universe (the one we see with our eyes and scientific instruments), and 2) the "supermundane" or intelligible universe – the one which is *ultimately* perceivable/comprehensible. The Traditionalist Frithjof Schuon wrote:

> One has to distinguish between "horizontal" and "vertical" dimensions, the vertical being supernatural and the horizontal natural; for the materialists, only the horizontal dimension exists, and that is why they cannot conceive of causes which operate vertically and which for that very reason are non-existent for them, like the vertical dimension itself.[2]

"Supernatural" does not refer to the paranormal (ghosts etc.), but to an order of natural that is simply beyond the ordinary order of natural and so is in fact still natural, just "super." And by "materialists" Schuon had in mind those who hold to the closed-minded intellectual position that is scientism. A mode of awareness can't be wrong, any more than a mode of physical conditioning can be wrong; it can, however, be less than ideal. The normal mode of awareness is less than ideal, for with it comes merely the conventional appreciation of a horizontal universe, which is not the full picture. It is when people adopt the intellectual position (which they needn't) that this conventionally appreciated universe *is* the full picture that we have a problem.

Higher Kingdoms of Nature

In this part of the book we are trying to get a handle on the cosmology of esotericism, which is a challenge as it features, for one thing, the idea that all astronomical bodies/objects are living sentient entities. And these LSEs are not biological *or* geophysiological entities (such as the Earth is in the Gaia hypothesis). The religious scholar Huston Smith claimed that this idea is the most difficult for the modern mind to grasp. This is because, born as we were into our modern age, our minds are firmly imprinted from childhood with the notion that there are three or four kingdoms of nature (or general kingdoms of being) at most – mineral, plant, animal, and, if we care to distinguish ourselves from the animal (which the life sciences struggle with, otherwise the rest of the time we do), human.

We can conceive of more advanced alien races – more advanced culturally and/or technologically (e.g. the Vulcans in *Star Trek*). But we struggle to conceive of entities/beings occupying kingdoms of nature *above* the mineral, plant, animal, and human. The medieval Christian mind would not rebel against this idea, for in that mindset there were angelic orders of beings above the human, with God at the top of the "Great Chain of Being." The ancient world also had its heroes, demigods, and gods between us and an ultimate God viewed as "the One." But the modern mind rebels against the idea that the human (the human-like on any planet, whether organic or cyborgic as the transhumanists imagine) is not at the top of the natural tree. Smith wrote:

> If things exist that are superior to us, they are not going to fit into our controlled experiments, any more than self-

> consciousness or advanced forms of abstract thinking would fit into (and therefore be brought to light by) experiments wood ticks hypothetically might devise.[3]

Both parts of the idea are very difficult for the modern mind to grasp. The first part is of entities which occupy kingdoms above us per se. OK, we might think, perhaps in an evolutionary future. But the idea is of such entities *already* existing — not "to be" but "already is." We commonly think of an evolutionary line from the Big Bang all the way up to us. We draw a horizontal line in our textbooks from particles on the left to humans on the right, tracing a supposed journey of billions of years from the simple and lesser in nature to the complex and greater. *This is the great horizontal myth,* the Traditionalist would say. Schuon wrote: "Transformist evolution offers a patent example of 'horizontality' ... owing to the fact it puts a biological evolution of 'ascending' degrees in place of a cosmogonic emanation of 'descending' degrees."[4]

The vertical alternative is from the greater to the lesser in nature: a spiritual *emanation* of levels of being rather than a material *evolution* just in/along one horizontal dimension. The vertical alternative is where the very existence of the greater level brings into existence the lower level, because the latter is being ideated by — or consciously experienced by — the former as its vehicle of expression (this is the nature of the "emanation"). As we proceed through this chapter and the rest of Part I, there are two analogies that will aid our understanding of this. The first is the author-character analogy.

Authors and Characters

An author imagines a character in a story. The character (let's call him Jack) also imagines a character in a story (let's call him David). So, David exists because of and in Jack (in Jack's imagination, as his "avatar" if we will), just as Jack exists because of and in the first author in the same way. There is a "first author" in esotericism — a greatest entity at the top of the vertical line, which we might call the Cosmic LSE. Then on the next level of being down, there are many lesser entities, which we might call sub-cosmic LSEs, which exist because of and in the Cosmic LSE. And on the next level down, there are many more yet lesser entities, which exist because of and in the sub-cosmic LSEs ... and so on down the line. Like a staff organization chart, with the ziggurat being an ancient symbol of this.

At some point down the line we reach a level of being on which there are solar system and planetary system LSEs. We (you and I, other humans, animals, plants, and minerals) exist because of and in one of those planetary LSEs — the Earth LSE — on the next level of being down again. This includes not only our physical bodies but also, with respect to humans and animals, our minds (our thinking and feeling natures). We exist because the Earth LSE does. We are being ideated or consciously experienced by It (the Earth LSE) as its vehicle of expression. The Earth LSE is analogously our "author" and we are its "characters." Collectively we are its avatar — or, as some might poetically put it, we are the Earth's dream.

Who Am I?

The second analogy is the physiological analogy. We are like cells in the body of the Earth LSE. Or better yet, we are like cells in the body of the *Sol* LSE, with the Earth LSE as an organ in that body (the Venus LSE, the Jupiter LSE etc. would then be other organs in the same body). The answer to the question "Who am I?" is analogously as a cell in the body of a "celestial divinity," with the direct appreciation of this subject to that supernormal mode of awareness called *henosis* in ancient Greek ("mystical union"). This is the transpersonal experience the person may have whereby it is as if a *greater entity* is looking out from behind the person's eyes at *Itself*. The person is a "participant" in this experience. On this subject, the Theosophist Alice Bailey wrote:

> First, the disciple becomes aware ... his consciousness is expanded until it might be called planetary consciousness. Secondly, he begins to merge that planetary awareness into something more synthetic still, and gradually develops the consciousness of the greater life [the solar] which includes the planetary life as man includes in his physical expression such living organisms as his heart or brain.[5]

The idea that you and I are not, in a sense, living ourselves, but are being lived by a greater entity – the Earth LSE – is startling, but perhaps not altogether unfamiliar. For there is our *immediate* perceptual experience to reckon with – before our modern, educated, rationalist minds kick in and tell us otherwise. A perceptual experience of living *in* an entity which is, of course, greater than ourselves. Esotericism challenges us to reflect on that immediate perceptual experience and to ask whether the modern idea that we small and imperfect beings are somehow "higher" on the scale of being than planets, stars, galaxies and so on is really that credible. Schuon wrote: "Modern science ... can describe our situation physically and approximately, but it can tell us absolutely nothing about our extra-spatial situation in the total and real universe."[6]

This "total and real universe" is the one featuring living sentient cosmic entities (including the one we live in) with the gist of the picture being something like this: The spatial size and majestic unity of astronomical bodies/objects, as we ordinarily perceive them, are but the self-representations in the horizontal universe of their own much greater extra-spatial size of being and (relative to us) unitive perfection in the vertical universe. This is why planets are, well, planets, and solar systems are solar systems, and galaxies are galaxies. Our appreciation error is a bit like, or would be a bit like, seeing the life of our local town and thinking this town level of life is the only level of life, when there is also, "above" this, the county level of life (including the county in which our town is located), and above this the state/province level of life (including the province in which our county is located), and above this the national level of life (including the nation in which our province is located), and above this the global level of life... with each level contained within the one(s) above and containing the one(s) below. Schuon continued: "Profane science, in seeking to pierce to its depths the mystery of the things that contain — space, time, matter, energy — *forgets the mystery of the things that are contained.*"[7]

We are contained in the Earth LSE (and wider Sol LSE). This includes our minds as well as our bodies — just as the character David's mind and body would be contained within Jack (and Jack's mind and body would be contained within his author). Another Traditionalist philosopher, Titus Burckhardt, wrote: "The rationalistic view forgets entirely that everything which it may express concerning the universe, remains a content of human consciousness."[8] This *human* consciousness then "resolves" into *planetary* consciousness, and finally into *solar* consciousness (as Bailey alludes to), for all along, just without us knowing it, we are being ideated by or consciously experienced by It (the Sol LSE) as its vehicle of expression. We are the Sun's dream beyond the Earth's dream. Relatedly, the third-century Neoplatonist Iamblichus wrote:

> An innate knowledge of the gods is co-existent with our very essence; and this knowledge is superior to all judgment ... and subsists prior to reason and demonstration. It is also co-united from the beginning with its proper cause, and is consubsistent with the essential tendency of the soul to *the good.*[9]

And fifteen hundred years later, the Platonist Thomas Taylor wrote:

> I confess I am wholly at a loss to conceive what could induce the moderns to controvert the dogma, that the stars and the whole world are animated, as it is an opinion of infinite antiquity, and is friendly to the most unperverted, spontaneous, and accurate conceptions of the human mind. Indeed, the rejection of it appears to me to be just as absurd as it would be in a maggot, if it were capable of syllogizing, to infer that man is a machine impelled by some external force when he walks, because it never saw any animated reptile so large.[10]

The Holarchical Universe

The universal structure is a *holarchy* (and we might bookmark this nature when it comes to the political model in Part II). This word brings together the words "hierarchy" and "holon." A hierarchy is a group of people (or any entities) organized into successive ranks, with each level being subordinate to (meaning both lesser than and dependent on) the one above. A holon is a thing which is simultaneously a whole and a part. The cosmic holarchy is of LSEs within LSEs, the lower rank of entities being dependent on the rank above *for their very existence*. And, as we can also analogize and think of cells within organs within bodies (or towns within counties within provinces), so we have a holarchy (and hierophany) and not just a hierarchy.

Behind the universal structure, beyond the Cosmic LSE at the top of the vertical line, is God (or a more religiously neutral term would be "the One"). "In order that Being may be brought about, the source must be no Being but Being's generator," wrote Plotinus.[11] God is not an LSE then (a Being), but contains within Itself, as the ultimate transcendent generator, the Cosmic LSE which, as the "first author," contains within itself all the other LSEs (including the Earth LSE and therefore us). The ranks of LSEs between God and us are, if we wish to call them such, secondary gods. It is a cliché used by modern historians that we invented the gods. No, says esotericism; in a very real cosmogonic sense they invent us (note the use of the present tense). As another Neoplatonist, Proclus, wrote:

> Let us as it were celebrate the first God, not as establishing the earth and the heavens, nor as giving subsistence to souls, and the generation of all animals; for he produced these indeed, but among the last of things; but prior to these, let us celebrate him as unfolding into light the whole intelligible and intellectual genus of gods.[12]

This genus of gods is the assembly of extraspatial LSEs. If the Earth LSE is like an organ, then all planets, in all solar systems, would be like organs with respect to the solar system bodies they are part of. And just as we humans may have that *henosis* experience, so too might other human-like beings on other planets. The idea of alien races that are culturally and/or technologically more advanced than us is a common one in science fiction. The idea of more advanced alien races in terms of self-awareness of their extra-spatial place in a universe of LSEs features in esotericism. This is an intriguing subject but is off-topic for this book.

What Am I Here For?

A different analogy would replace "organ" with "business department," and "cell" with "worker." A department that is fully integrated into the wider organization but has its own specific function to perform. The answer to the question "What am I here for?" is to consciously cooperate in our planet's function as a "department" of a "local office" (our solar system) of a vast cosmic "organization." "Cooperating in" means helping to make it the best-performing organization it can be — which means creating and sustaining, in a continuous improvement manner, a sublime (beautiful, just, and peaceful) planetary civilization. To do so would be to express the Good — which is the universal purpose — and our ultimate felicity depends upon this (as we can imagine it would).

The message coming especially from this different analogy is that the task of successfully running our planet has, in a way yet to be appreciated, a wider "business context." This being the case, the younger generation of today and their successors, who particularly have the task of reaching globalization's destination (planetary government), should not just look "horizontally outwards" at our political management task set within conventional science's spatial positioning of the Earth, but also look "vertically inwards" at our purpose-in-existing business task set within esotericism's extra-spatial positioning of the Earth. The twentieth-century esotericist Manly Hall, in typically pithy words, summed up the picture and the task thus:

> The gods are modes of universal consciousness, that is, they are degrees of awareness in space. Humanity is basically also a degree of awareness, and so are all the other kingdoms of Nature ... [The task is to] bring the kingdoms of the earth into harmonic concord with the kingdom of space.[13]

Chapter 2

The Nature of Things

Introduction

Contrary to what we've been taught, the universe is made of consciousness/life, not matter/energy. Looking from the bottom upwards, it is like one level of characters being imagined by authors on a level above, who in turn are just characters being imagined by authors on a level above, and so on up to a first author at the top and an incomprehensible generating principle (God/the One) behind that. Put another way, the universe is but a dream within a dream as the Bard said. There is no *matter* existing outside of (independently of) any dream, any dreamer. The Theosophist Helena Blavatsky wrote: "From the stand-point of the highest metaphysics, the whole Universe, gods included, is an illusion; but the illusion of him who is in himself an illusion differs on every plane of consciousness."[1] And: "The Universe is in reality but a huge aggregation of states of consciousness."[2]

Films such as *The Matrix* resonate with this esoteric reality, although we are not literally talking about simulated realities; the substance of reality is "consciousness-stuff," not matter/energy of any kind. And there are multiple levels of being, not just the one. In *henosis,* the person appreciates his/her "character-nature" and, concomitantly, the character-nature of everyone/everything else. We are all, in one sense, being lived by a greater entity above. *"Mitakuye Oyasin,"* as the Lakota people of North America say ("we are all one family"). We are all one family in the Earth LSE. It is our "sky father" married, if we like, to the "earth mother" of our sense-perceived planet.

Spheres within Spheres

> For the Deity, intending to make this world like the fairest and most perfect of intelligible beings, framed one visible animal comprehending within itself all other animals of a kindred nature ... Now to the animal which was to comprehend all animals, that figure was suitable which comprehends within itself all other figures. Wherefore he made the world in the form of a globe.
>
> — Plato[3]

A level of being is a sphere in esotericism. So too is an LSE. The conscious experience of any entity extends out to a certain "circumference" around the self at the center of the experience. At the top level is the Cosmic LSE (the one *ultimately* visible i.e. perceivable/comprehensible animal in the quote from Plato above). It has the form of a globe (a sphere). It comprehends within itself (as an author does his/her imagined characters) all other animals of a kindred nature (also authors at the same time as being characters). The circumference of any sphere corresponds to the range of conscious experience of the entity, or in the case of a level of being, all the entities making up that level. The whole universe can thus be pictured as a kind of Russian doll of spheres within spheres. At the center of all the spheres — although not as a "dot" because it is not an LSE but contains within Itself all LSEs — is God/the One. As the anonymous Hermetic text *Liber XXIV Philosophorum* puts it:

> God is an infinite sphere whose centre is everywhere and whose circumference is nowhere ... God is beginning without beginning, unchanging progress, endless end ... God is what alone lives from its own intellection ... God is the darkness remaining in the soul after every light.[4]

The Consciousness Cycle

This pertains also to the above quote. A person wakes up in the morning and is conscious. He goes to sleep at night and is unconscious. The cycle repeats. We may discern three aspects to consciousness when it is "on" (i.e. when the person is awake): 1) there is a self that is conscious, 2) there is the content of that self's consciousness (what s/he consciously experiences), and 3) there is the activity itself — the activity of consciously experiencing — going on. The Cosmic LSE is the biggest consciously experiencing entity. When it is "awake" there is: 1) a self that is conscious, 2), the content of that self's consciousness — *which is the entire universe*, and 3) the activity of consciously experiencing going on. When the Cosmic LSE is not awake, there are none of these three aspects or "divine principles" manifesting. Consciousness is not on, therefore there is no universe.

In esoteric mathematics, there is a sequence to the divine principles. A self that is conscious is the number 1 (in Greek, the *monad*). From this comes the number 2 (the *dyad*), as we have here now a principle which is really two things — for you can't have the content of a self's consciousness without a self. From this comes the number 3 (the *triad*), as we have here now a principle which is really three things — for you can't have the activity of consciously experiencing going on without the other two. From this comes the number 4 (the *tetrad*), as we have here now... well, life as we know it — for you can't have that without the three principles behind it. 1+2+3+4=10 (the all-inclusive *decad* and the all-important *tetractys* of the Pythagoreans):

> I swear by him who the tetractys found,
> And to our race reveal'd; the cause and root,
> And fount of ever-flowing Nature.
>
> — The Pythagorean oath from *The Life of Pythagoras*[5]

The idea of cycles is a common one in the world's religions and especially in Hinduism which has a strong relation to Western esotericism. When consciousness is on, there is "light." When it is off, there is "darkness." We recall the God in Proclus's quote "unfolding into light" the tiers of gods. That would be the cosmogonic emanation — the unfolding of the levels of LSEs as consciously experiencing entities. The Great Cycle of the universe has no beginning and no end — it goes on "day after day." Blavatsky wrote: "We believe in no [one-off] *creation*, but in the periodical and consecutive appearances of the universe from the subjective on to the objective plane of being."[6]

With each Great Cycle there is progress. It is like a worker picking up his tools again in the morning and furthering his work building a glorious temple (something expressing the Good). The worker is the Cosmic LSE, within which are lesser workers on all the levels. At our own humble level, and in our own limited capacity, we are to be such workers. A human being is also a living sentient entity. There is a level of being below us (referring now to the strictly corporeal). Each of us is a greater entity containing a myriad of lesser entities. We are "Jacks" to a host of "Davids." They — the myriad of lesser entities making up our observed corporeal bodies — are collectively our avatars, just as we humans, animals, plants, and minerals are collectively the Earth LSE's avatar. These lesser entities are, in their essential form, spheres. And so are we.

Esotericism, Metaphysics, and Science

All LSEs, all spheres, have an internal dynamic of their own and they interact. At the level below us, these lives (invisible to conventional science as they are nonmaterial) interact, just as LSEs on every level do. We read the metaphysical thought of ancient philosophers as pre-scientific, that is, struggling with ideas that we firmly grasp now about such things as the atom and gravity. But in truth, the atom the esotericists among the ancient philosophers were talking about was the consciously experiencing entity sphere. And the forces they were talking about were those relating to their internal dynamism and interactivity, including combining into the corporeal forms we see. Thus, as Blavatsky said: "Modern science is ancient thought distorted, and no more."[7]

This distortedness applies to the idea that the universe is made up of atoms when it is made up of lives, and to the idea of biological evolution when there is, rather, a consciousness progression through world periods (see next chapter). When it comes to Plato's *Ideas*, these are the universals manifested on all the levels of being. There is Beauty for instance in corporeal nature (as any gardener, artist, or physician knows). There is Beauty in the macro-world too, which is the level above the corporeal and our level of being proper — manifested in societies that live in harmony with nature. And there is Beauty at all levels above. Other Ideas include Justice and Love. We are to understand that the higher the level of being, which is also to say the nearer to God at the top it is, the more qualitatively superior it is. In other words, the more sublime or perfect existence is on those levels. Few would argue that existence on our level is far from perfect.

Along the same lines as Blavatsky's physics-is-but-ancient-thought-distorted message, a few decades later the Traditionalist René Guenon described how, for the ancient philosophers, physics was secondary to or a subset of metaphysics. Physics was the study of nature, but the *whole* of nature included the vertical dimension. One error of modern humans — or as Guenon put it, one part of the intellectual decline in the West — was the division of physics into discrete natural sciences. A much bigger error was the separation of physics from metaphysics, such that all we had left now was "profane science" rather than "sacred science" (sacred science being the science of the whole of nature). Today, the idea of sacred science is scoffed at by modern science, whereas in the past, profane science had to justify its existence and its worth in relation to sacred science. Guenon wrote:

> In seeking completely to sever the connection between the sciences and any higher principles, on the pretext of safeguarding their independence, the modern conception robs them of all deeper meaning and even of any real interest from the point of view of knowledge, and it can only lead them down a blind alley, imprisoning them, as it does, within an incurably limited realm [the horizontal]. Moreover, the development which goes on inside that realm is not a deepening of knowledge, as is commonly supposed; on the contrary, the information so gained remains superficial and consists merely in that dispersion in detail that we have already alluded to, in an analysis as barren as it is laborious and which can be pursued indefinitely without advancing a single step further in the direction of true knowledge. Furthermore it is not for its own sake that Westerners in general cultivate science as they understand it; their primary aim is not knowledge, even of an inferior order, but practical applications, as may be inferred from the ease with which the majority of our contemporaries confuse science and industry, so that by many the engineer is looked upon as a typical man of science ... In assuming its modern form science has not only lost its depth, but also, one might say, its solidity, since attachment to the principles enabled it to participate in their immutability to the full extent that the nature of its subject matter allowed; once shut off exclusively in the realm of change, however, it cannot hope to achieve any kind of stability, nor to find any solid basis on which to build; no longer starting out from any certainty, it finds itself reduced to probabilities and approximations, or to

> purely hypothetical constructions which are merely the product of individual fantasy ... Modern science, arising out of an arbitrary limitation of knowledge within a certain particular order which is indeed the most inferior of all, namely that of material or sensible reality, has as a consequence forfeited all intellectual value, so long that is to say as one uses the word intellectuality in all the fullness of its true meaning and refuses to participate in the "rationalist" error, or to reject intellectual intuition, which amounts to the same thing.[8]

Guenon was a fierce critic of modernity in general but would have agreed with Schuon and others that it is scientism, really, that is the problem, not science per se (for that is just the pursuit of knowledge – in what dimension one looks for it and with what mental faculty is a different matter). Back in the eighteenth century, Thomas Taylor expressed the same concern:

> The discoveries of experimental philosophy [modern science], float like straws on the surface, while the wisdom of Pythagoras and Plato lies concealed in the depths of the river ... Is it to be supposed, that in an age when philosophy was almost adored; when it was esteemed by kings, cultivated by noblemen, and even reverenced by the vulgar; when empire was relinquished for its pursuit, and every danger encountered for its possession: is it to be supposed, that nothing but delusion was the offspring of so glorious a period, and nothing but folly the reward of such generous endurance? Or shall we say, that the discovery of truth was reserved for the age of experiment; and that she is alone to be apprehended in the infinite labyrinth of particulars? That she is to be investigated with the corporeal senses, and not with the powers of intellect; and that the crucible, the alembic,

> and the air-pump are the only mediums of detection? ... Shall we call this the age of philosophy, in which talents are prostituted for sustenance, and learning submits to the impudence of wealth? ... Again, the object of the Pythagorean and Platonic philosophy was to make its possessors wise and virtuous ... but the object of modern philosophy, is a promotion of the conveniences and refinements of life, by enlarging the boundaries of traffic; and the mathematical sciences are studied solely with a view to this enlargement.[9]

We can detect an intellectual decline in the West with regard to metaphysics (although in recent years there has been the beginnings of a revival). On the other hand, there were very few intellectuals in the past, and intellectuals are in large part our cultural leaders. Today's intellectuals have simply developed their rational-scientific minds rather than, or more than, their "intellectual intuitions." This problem of rationalism, such as it is, distinguishes modern Western culture, and is an educational issue. Perhaps we might call conventional Western scientific knowledge "Type 2 knowledge." This would still honor it as knowledge, while reminding us that there is a presiding "Type 1" sort. As a cultural activity and as a definition, science need not be confined to: a) the sensible/horizontal universe, and b) study through reason and experiment. That we are aware of an intellectual straitjacket when it comes to the scientific method should be all we need to let us know that there must also be a Type 1 knowledge.

We need a philosophical education in the sense known by Plato and Pythagoras. Then, with *both* mental faculties developed (the rational mind and the intellectual intuition — the lesser intelligence and the greater intelligence), we can do more good than was possible in their day, having more cultural leaders, that is, more wise and virtuous people helping to build that beautiful, just, and peaceful planetary civilization. Philosophy is the love of wisdom, and wisdom, Pythagoras reminded his students, is the science of the truth that is *in* beings. Study that science, which is to say study consciousness (which requires a methodology consistent with that project), and the universe made of consciousness — the ultimately comprehensible universe — will reveal itself. Hence the Delphic maxim "Know Thyself." Part of that knowledge is of our consciousness origins in a greater entity above. As Alice Bailey wrote in 1925:

> The confines of the Heavens Themselves are illimitable and utterly unknown ... Go out on some clear starlit night and seek to realise that in the many thousands of suns and constellations visible to the unaided eye of man, and in the tens of millions which the modern telescope reveals there is seen the physical manifestation of as many millions of intelligent existences ... Realise further that the bodies of all these sentient intelligent cosmic, solar and planetary Logoi are constituted of living sentient beings, and the brain reels, and the mind draws back in dismay before such a staggering concept. Yet so it is.[10]

The brain reels at the *mysterium tremendum et fascinans* ("fearful and fascinating mystery") of a universe made of consciousness/life, not matter/energy. Esotericism would have us appreciate: 1) that conventional Western science, in its dealings with corporeal nature, is somewhat of a blind bull in a priceless china shop — not really knowing what it is dealing with (which is lives); and 2) that as we look to develop a more holistic and global worldview, a continued "horizontality" only, however less reductionist, would still be a limited and Western-only view. If we are to have a truly global worldview, inclusive also of the wisdom of the East and that of many indigenous peoples, "verticality" would need to make its reappearance.

Another aspect of the universe of consciousness is what Huston Smith referred to when he wrote of the hierarchy of levels being like a pyramid of magnets, with those lives on each level or tier being "attracted to the tier above while being empowered by that tier to attract the magnets below them."[11] This attraction is a *psychological* attraction — entities on one level being attracted to the level above, on account of having an intuition that their essential selves have a home there and ultimately return to it. This "Great Return" is the subject of the next chapter, and previewing this we might include the following from the second-century Gnostic text *The Gospel of Truth,* attributed to Valentinus:

> Each one will speak concerning the place from which he has come forth, and to the region from which he received his essential being, he will hasten to return once again. And he want from that place — the place where he was — because he tasted of that place, as he was nourished and grew. And his own place of rest is his Pleroma. All the emanations from the Father, therefore, are Pleromas, and all his emanations have their roots in the one who caused them all to grow from himself. He appointed a limit. They, then, became manifest individually in order that they might be in their own thought, for that place to which they extend their thoughts is their root, which lifts them upward through all heights to the Father.[12]

Chapter 3

Past-Present-Future

Introduction

Conventional science posits a universe that began 13.8 billion years ago with a Big Bang. As for when it will end, there is no agreed theory, but a recent one is the Big Rip, whereby as the universe expands it rips itself apart — we perhaps only have 22 billion years left. The most popular theory, though, is that everything just peters out in trillions of years' time. There may or may not have been other universes before this one and afterwards (most scientists subscribe to a "one universe only" view), but one or many, the beginning, middle, and end of all universes is but matter/energy and there is no inherent purpose to any of them. This is the averred backdrop to human life, relativizing everything we think and do (including how we may run our planet) as a massive waste of time and energy. This is utter nonsense, says esotericism. It's just the story modern Western humans tell themselves around a campfire.

Esotericism posits a "day after day" universe of the nature described in the last chapter — cosmic consciousness on, cosmic consciousness off — with no first or last day, and with each day being a progression. With cosmic consciousness on, there are three dimensions of time, which in the author's previous book were called Character-time, Aevertinity and Consciousness-time. "Character-time" is ordinary time. This flows from left to right, where yesterday is the past, today the present, and tomorrow the future. Even if we could time-jump back to yesterday, or time-jump ahead to tomorrow, wherever we "land" we would immediately proceed to pick up the same left-to-right, forward-marching stream of Character-time. This dimension is a one-way arrow and is the ordinary time we experience and measure.

Aevertinity

"Aevertinity" is not the same as endless duration of ordinary time, for it is outside ordinary time altogether. Returning to the author-character analogy, as far as the character Jack is concerned (the same would be true for David), he is real and so is the world he lives in — he does not realize both he and his story-world are being imagined by an author. Jack lives in Character-time — which is to say, he has experience after experience, birthday after birthday, for however long Jack lives. Even if Jack lived for billions of years, there would still be the outside time dimension relating to his world, only existing in an author's imagination and therefore *not in the time of his world*. Aevertinity is the ever-present "now" that is beyond ordinary time. Aldous Huxley referred to it in his classic work *The Perennial Philosophy*:

> The manifold world of our everyday experience is real with a relative reality that is, on its own level, unquestionable, but this relative reality has its being within and because of the absolute Reality ... [This] divine ground of all existence is not merely a continuum, it is also out of time.[1]

Consciousness-Time

"Consciousness-time" is the third dimension of time. Jack's purpose is to "wake up" and appreciate that he is not living himself but is being lived, in one sense, by an author above. As part of this he appreciates that the world which he took as real is not real (at least not in the sense that he took it to be), and that he and all the other characters in his world were brought into being and are likewise to wake up. So, there is a dimension of time here involving an outward creational arc (that brought Jack into being as the objectively real person he thought he was), and then the inward psychological arc that takes Jack out of being himself into conscious union with his author. Consciousness-time is thus symbolized by the circle. Ordinary time is symbolized by the horizontal line. And Aevertinity is symbolized by the *axis mundi*. All three together would be symbolized by the circle with the cross in it (see Figure 1).

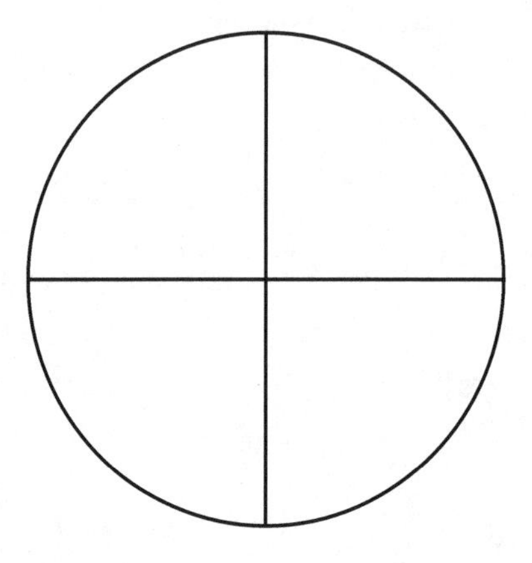

Figure 1: The Solar Cross

What is truly *Eternal* is the repeating of the Great Cycle — but there is a progression each time so it's not mere Groundhog Day. With every Great Cycle, which is to say with every universe, there is: 1) the Aevertinal or ever-present "pole"; 2) the outward or coming-into-being arc, and the inward or coming-out-of-being arc, which takes place in Consciousness-time; and 3) ordinary or Character-time which, though it seems like ultimate and exclusive reality to the characters in the story-world, is but relative to a particular mode of awareness. *The one-way arrow from a Big Bang to a final universal end is part of the great horizontal myth*. Or to put it another way, this is only reality when we haven't got our esoteric glasses on. This means there's no justification for nihilism or apathy. On the contrary, we can be life-affirming and we have a job to do, as referred to in Chapter 1.

"Permeating occidental cognitive orientations are concepts of time conceived as lineal, divided into mutually exclusive segments of past, present, and future," wrote the religious scholar Joseph Epes Brown.[2] Native American experience of time, by contrast — and this is common among indigenous peoples — tended to be in terms of the circle or cycle (Consciousness-time), with an orientation to the "unchanging mystery of the present moment of a Now beyond Time" (the Aevertinal).[3] The cultural historian Mircea Eliade described how modern, non-religious humans live in a desacralized cosmos both in terms of space and time. In relation to space, and referring to our place in the universe, there is no "qualitative differentiation and, hence, no orientation [is] given by virtue of its inherent structure."[4] In relation to time, humankind no longer participates in a yearly rebirth modeled on the Great Cycle following the Great Return.

This desacralized cosmos is the cause of our existential suffering and longing for meaning. But as Huston Smith said, "a meaningful life is not finally possible in a meaningless world."[5] A meaningless world is the purposeless universe presented by conventional science. This is why the cultural historian Richard Tarnas wrote that: "No amount of revisioning philosophy or psychology, science or religion, can forge a new worldview without a radical shift at the cosmological level."[6] A radical shift would include the reappearance of verticality, and in the extraspatial universe, or with respect to that, the recognition that there are these three (or four if we include the Eternal) temporal dimensions.

Cosmic consciousness is currently on — we're in a universe right now — thus there is an Aevertinal pole along which we might picture, like a hanging necklace of pearls (and as a complementary image to the "spheres within spheres" picture from Chapter 2), the levels of being and the LSEs on them. This pole is the ever-present temporal center of our experience. It is with respect to this pole that it can be said the universe did not begin in the past but in the present — *referring not to the present of ordinary time, but to the ever-present pole*. This is not the easiest idea to grasp, admittedly, but the author-character analogy helps. Brown also described how, for the Lakota people, "time has a *spatial extension*: that which once happened literally took place, and *still has a place*."[7] This relates to the subject of world periods covered in the next section.

A Great Cycle is complete (and is followed by "sleep" and no universe, prior to a new day and a new universe) when the return journey is completed: when all the characters have "woken up." At the largest scale, there is an outward or coming-into-being arc that brings the whole cosmos (all the levels of being, and all the entities on those levels — collectively the universal "Jack") into being, and then an inward or coming-out-of-being arc that takes the whole cosmos out of being. Our level of being is one of those pictured "pearls." It is the only one we are really interested in, as it is the only one we can really appreciate. It is only our level of being and, indeed, our solar system that we are really interested in — which connects with our ultimate "avataric" relationship with the Sol LSE.

We all have to wake up. By "all" is meant all humans ... but this also refers to all animals, all plants, all minerals. *Mitakuye Oyasin*: they are all our "younger brothers and sisters" who have not yet attained the type or modality of consciousness we have, ahead of them, which is a prerequisite for starting the second leg of the journey. That which once happened "still has a place" as we look around us, in a contemplative way, at the lower kingdoms of nature. Relatedly, the esotericist Rudolf Steiner wrote:

> To a man who can look with understanding at the spiritual that is there now, hidden within what is manifest to the senses, insight also into former states ... however distant,

> cannot appear essentially impossible... [O]nce we are able to recognize the presence of the spiritual here and now, we shall find the earlier conditions given or implied in the immediate vision of the present, just as the condition of the one-year-old infant is implied in the appearance of a man of fifty.[8]

World Periods

Esotericism speaks of seven world periods spanning the outward and inward arcs of Consciousness-time. We are in the fourth world period (of our local Sol system). Periods 1–4 see, respectively, each of the four kingdoms come into being — mineral, plant, animal, and human. This would refer to kingdoms *in terms of types of consciousness*. And in the end, these are just subsets of one kingdom/type of consciousness. So, more accurately, periods 1–4 see, respectively, more mature forms of the one kingdom come into being. Period 4 sees the most mature form of this one kingdom (us) come into being, but period 4 also sees humans pass out of being as we wake up. We can read the coming in of human beings (the human type of consciousness on the Earth, but this would also apply to human-like beings on any planet — the world periods relating to our common level of being or to the conventionally appreciated horizontal universe) in the *Metamorphoses* of the Roman poet Ovid:

> A creature of a more exalted kind
> Was wanting yet, and then was Man design'd:
> Conscious of thought, of more capacious breast,
> For empire form'd, and fit to rule the rest:
> Whether with particles of heav'nly fire
> The God of Nature did his soul inspire,
> Or Earth, but new divided from the sky,
> And, pliant, still retain'd th'aetherial energy:
> Which wise Prometheus temper'd into paste,

> And, mixt with living streams, the godlike image cast.
> Thus, while the mute creation downward bend
> Their sight, and to their earthly mother tend,
> Man looks aloft; and with erected eyes
> Beholds his own hereditary skies.
> From such rude principles our form began;
> And earth was metamorphos'd into Man.[9]

As for periods 5, 6 and 7: Period 4 animals (already having *something* of the human in them in terms of consciousness – which is why a person can be so close to their cat or dog for instance) are, in period 5, humans (what we are today – we who already have something of the *supra*human in us, hence our inherent religiosity even if that takes the form of believing in secular humanism), and they wake up during that period. Period 4 plants (already having something of the animal in them, hence their sensitivity as gardeners know) are animals in period 5, and period 4 minerals (already having something of the plant in them, even though we struggle to apprehend it) are plants in period 5. Then in period 6 everything shunts along again. What were plants in period 4 and animals in period 5 are humans in period 6 and wake up. What were minerals in period 4 and plants in period 5 are animals in period 6. Finally, there is only one subset kingdom or consciousness in period 7 – humans that were minerals in period 4, plants in period 5, and animals in period 6. Again, it is important to recall we are talking about types of consciousness and, for most intents and purposes, we don't need to consider the periods prior to (1–3) and post our own (5–7). Period 4 is where it is all happening, so to speak.

Periods 1–4 are also called, in different language, Fire, Air, Water, and Earth. What has taken place up to now (the Earth period) is a "condensation" in the mind of the celestial divinity concerned (the Sol LSE) of its "idea." The nebular hypothesis is a distorted reflection of this esoteric solar genesis. Any idea starts out as a vague and homogeneous *something*, and finally condenses into a definite and heterogeneous *thing*, and these positions have a correspondence with the mineral and human types of consciousness – the first being barely conscious of anything at all (or nothing at all as we might adjudge), and the last being as we everyday experience it. The earlier periods, looking at our "kin," can therefore be seen in the present, "hidden within what is manifest to the senses."

We are currently in the latter half of the fourth period, and that period up to now can also be seen in the present. The fourth period is that which sees the human type of consciousness first come into being. Around us we see human beings occupying different age groups, and our present esoteric historical age within the fourth period is the "young adult" age (or at least that's what the author proposes to call it – the Theosophist might call it the fifth root race). Before this age were three or four ages, depending on whether the first is counted, which we could call the "prenatal," "infant," "child," and "youth" ages. The next age in the sequence is the "mature adult" age. We will consider these ages especially as they relate to socio-economic systems in Chapter 5.

The new age to come (or actually already begun – it's just that there is a crossover) would be mere millenarianism if it didn't belong in a natural sequence of consciousness, and our free will enables us to fast-track or slow-track the fullness of this age but not derail it (just as we can go along with or resist growing up, but cannot escape it). In a wiser age ahead we can therefore be assured, but not complacent. It is in the interests of all of us that we do fast-track to the fullness of the mature adult age or, to use a sporting phrase, get our heads in the game. This is obviously a different vision of the future than that postulated by scientific-atheist writers who see only more and more mechanistic thinking and, in parallel, more and more mechanization of the human body. This book strongly contends it is the transhumanists who are the fantasists, though, not the esotericists.

We naturally ask: What is the relationship between the world periods and ordinary (exoteric) history? The answer requires us to appreciate — in just the same way as with science — a distinction between profane and sacred. Sacred history would be the history of the *whole* of nature (a Type 1 knowledge), as sacred science is the study of the whole of nature. In the modern West, we have dropped metahistory as we have dropped metaphysics. We can imagine Jack asking a fellow character in his story-world: "How many billions of years ago did the periods start?" And that character, equivalent to an esotericist in his story-world, replying: "You're missing the point, because these periods trace the coming into being of the dream you're living in."

If we insist that the horizontal universe is objectively real (which is not scientifically verifiable), then the past must be measurable in ordinary time. But we do not need to insist on this, any more than we need to insist that the cosmic holarchy is spatially measurable. The Traditionalist would say that to insist on such would, in fact, be intellectual laziness. For that would merely be to follow the way the rational mind works — a way that requires things to be fixed and to have finitude. But if we follow the way our other mental faculty works (the intellectual intuition), there is no problem in accepting that the horizontal universe is only *relatively* real, to the extent that we can give primacy not to ordinary linear history but to the time dimension that brought this relative reality into being (Consciousness-time). The future and the past, as we give our minds over to contemplating them, can then be framed within the symbol of the circle with the cross in it.

> The basic error of systematized rationality ... is to put fallible reasoning in place of infallible intellection; as if

> the rational faculty were the whole of Intelligence and even the only Intelligence.
>
> — Frithjof Schuon[10]

Intellection is the action or process of understanding; the error is when we think this exclusively means reasoning (what the rational faculty does), as Schuon says above. Our subject in this chapter, put into the form of a slogan, is: "It is the history of consciousness that counts." Especially as our consciousness, in one sense, isn't ours anyway. We have spoken about "waking up" and thus employed an eschatological or soteriological phrase. But this business is not chiefly about personal enlightenment or gnostic salvation, but about service to and within the cosmos. Bailey wrote:

> Men [i.e. human beings] are apt to think that the whole evolutionary process — including the development of the subhuman kingdoms in nature — is merely a mode whereby men can reach perfection and develop better forms through which to manifest that perfection. But in the last analysis, human progress is purely relative and incidental. The factor of supreme importance is the ability of the planetary Logos to carry out His primary intention and bring His "project" to a sound consummation, thus fulfilling the task given to Him by His great superior, the Solar Logos.[11]

How Am I to View the Universe?

By way of concluding Part I, the answer is: In the opposite of the materialist, reductionist, mechanistic way. Plotinus wrote in the third century in words that are just as apposite today: "Those to whom existence comes about by chance and automatic action and is held together by material forces have drifted far from God and the concept of unity."[12] Have drifted far from appreciating the ultimately comprehensible (horizontal plus vertical) universe, that is. When some physicists today declare they have a near-complete knowledge of everything, esotericism tells them to take a cold shower – they are not on the threshold of omniscience, but of humility in the face of a vertical universe that is beyond their ken and awaits their discovery.

Is there a God? Yes, although we may prefer a more religiously neutral term such as "the One" in light of it not being a Being, but Being's generator. This generating nature allows us to be nontheistic if we so choose. Is the universe made of matter/energy? No, it's made of consciousness/life – a "dream within a dream" as the Bard said. Did the universe begin *x* years ago and will it end in *x* years' time? No, there is cycle after cycle of existence, with no beginning and no end, each time an opportunity and with purpose to express the Good. We can be life-affirming and optimistic, and have a specific and noble task to perform, which is to create a beautiful, just, and peaceful planetary civilization. Deep down we know this is the case – do we not?

Planets, stars, galaxies ... all are living conscious entities contained within larger entities of similar kind, and containing within themselves lesser entities of similar kind. A person's felicity – also that of a planet, a star, or a galaxy – lies in its self-perfection, or the maximizing of its creative potential in terms of such built-in Ideas as Beauty, Justice, and Love. These Ideas are collectively the Platonic "Idea of the Good." There is no material cause of anything – chemistry comes from life, not the other way around; and the brain (the corporeal body generally) comes from consciousness, not the other way around. There is not one level of being but many, all existing in the Now beyond Time. To the specific question of how one should look at the ordinary big picture of conventional science compared with the grander big picture of esotericism/the perennial philosophy, Huston Smith answered thus:

> The perennial philosophy upstages the best show science can manage. For without backing off in the slightest on numbers, it makes, as it were, a right angle turn in to a wholly new dimension: that of quality-qualitative experience [the vertical] ... To see what this involves, we might try to imagine the qualitative difference between the experience of a wood tick and ourselves, and then, continuing on in the same expanding direction, introduce orders of magnitude that science has accustomed to us: 10^{23} or whatever ... If our imaginations could concretely effect such moves we would have no difficulty understanding Plato's exclamation: "First a shudder runs through me, and then the old awe creeps over me."[13]

The old awe refers to the *mysterium tremendum et fascinans*. The postmodern philosopher is partly right when he or she says there are only truth-claims, but as we must include this truth-claim (that there are only truth-claims) in with all the others, then we are not obliged to cut our mental coats according to this ultimately unsatisfying and not-going-anywhere pluralist and relativist cloth. Furthermore, what this person fails to appreciate is that *all thought is contained within consciousness* — and that consciousness belongs to the extraspatial universe (in the first instance, to the Earth LSE). We are all one family with respect to that greater life: this is no mere romantic sentiment or Pagan/Neopagan belief but a fact. We finish Part I in the same place as we started — with our place in the universe. As Seyyed Hossein Nasr said, *it all comes down to consciousness*:

> Were we to accept the truth that "in the beginning was consciousness" ... we would no longer feel as aliens in a dead and forbidding cosmos, as accidents in a lifeless universe. Far from being aliens, we would feel once again at home in the cosmos as did traditional men and

> women over the ages. Our rapport with animals, plants and even the inanimate world would change from one of strife and need for control and domination to one of harmony and equilibrium ... Our deepest values, our attraction to and yearning for beauty, peace and justice, and the experience of love itself on all levels would not be seen as being simply subjective states devoid of any objective reality but on the contrary as corresponding to cosmic and ultimately metacosmic realities. And our ethical actions and norms, far from being simply based on standards set by merely human decisions and agreements, would be seen as having a divine origin and cosmic correspondences and as being much more real than simply convenient accommodations created by human societies for their survival or selfish interests.[14]

Coda

Before moving on to Part II, we might form a bridge here between the two parts of the book by rewinding to the point made about the postmodern philosopher who says there are only truth-claims, but has the blind spot of not seeing that this must also apply to the truth-claim that there are only truth-claims. The *post*-postmodern philosopher does not have this blind spot, but as a result, when asked what the truth is then, is silent. Into this "post-truth" silence arise competing voices of all kinds — the internet being both a vehicle for and a symbol of this.

In response to this post-truth cultural condition, we can understand why there is a reactionary philosophical conservatism in the form of either scientism or religionism, and a reactionary political conservatism in the West in the form of Anglo-Saxon capitalism, nation-centrism, and "anti-wokeism." Western culture thus appears "stuck" in an unsatisfactory and deleterious mental state, where a background materialist worldview still dominates (the one which says that the universe is all just matter, pointless, and will end in oblivion), and our leading public intellectuals, while not able to subscribe to this worldview entirely, have little or nothing substantively to say in its place (and retreat rather into the postmodern mindset and cul-de-sac).

Western culture, now globally dominant, appears also stuck in an unsatisfactory and deleterious political state, where a background capitalist world system still dominates, and while our leading progressives are not able to subscribe to this system in whole or in part, struggle to offer an alternative that is genuinely *beyond* either capitalism or socialism. Part II is, in a sense, an attempt to fix this problem. It includes, as we shall see, a political model (rather than an ideology) that reflects the holarchical nature of the universe. It includes, as we shall see, an appreciation that economic systems do not produce societal consciousness but rather the other way around. We are, today, in between one societal consciousness producing a capitalist world system, and another societal consciousness that will produce a world system which is not based on wealth at all (whether in private or public ownership) but on wellbeing. It includes, as we shall see, a realization that what will save the liberal rules-based international order currently under threat from illiberalism, authoritarianism, and nationalism, is not constraining the international and releasing the liberal (as the Right would have it), but constraining the liberal (so businesses have to operate in a socially and environmentally responsible manner) and releasing the international (so proper and actual global governance can arise through or alongside the United Nations).

Part II
Global Solutions

Chapter 4

Good Global Government

Introduction

The big picture of the universe is one thing. It behoves us also to recognize the living conditions of large parts of humankind today. According to the United Nations, approximately 1 in 8 people live in extreme poverty, 1 in 9 suffer from hunger (even in the wealthiest country in the world — the USA — 1 in 8 people struggle with hunger), 1 in 3 are malnourished, 1 in 4 children under the age of five have stunted growth, 1 in 11 have no option but to use unclean water, 1 in 7 are without electricity, 1 in 8 live in urban slums, and 1 in 2 work in vulnerable conditions. That the figures on *extreme* poverty were worse in recent decades is true. That nations are officially signed up to the 2030 Agenda with its 17 goals, chief among which is to eradicate poverty "in all its forms," is also true. That we will struggle to achieve these goals (as of 2023, only 15% of its targets were on track to being met) ... well, that can be assumed, with most global thinkers accepting that poverty eradication under the current economic system would take centuries to achieve if it was possible at all under that system (which, as it created the problem in the first place, is not credible).

Recalling Paul Kennedy in the Introduction, if we were visited by that extraterrestrial we would likely feel ashamed — and rightly so — at the current human situation (to say nothing about the current environmental situation, which is more than adequately covered by other books). What that extraterrestrial might point out is that: a) we have the technical means to end this situation, and b) we signed up in 1948 to doing so — or so it would appear, referring to the agreed human *entitlements* contained in the Universal Declaration of Human Rights. The extraterrestrial might even quote back to us the words of the economist E. F. Schumacher:

> The generosity of the Earth allows us to feed all mankind; we know enough about ecology to keep the Earth a healthy place; there is enough room on the Earth, and there are enough materials so that everybody can have adequate shelter; we are quite competent enough to produce sufficient supplies of necessities so that no one need live in misery ... There is no economic problem and, in a sense, there never has been. But there is a moral problem ...[1]

The Purpose of Government

There is no perfect "off-the-shelf" political philosophy any more than there is a perfect off-the-shelf natural philosophy, but behind all human thought (contained as it is within consciousness) is the universe, and the way it governs itself, so to the extent that we wake up and appreciate that total and real universe, we can put into effect a civilization that is reflective of it. "As above, so below" is the esoteric axiom; that is, as the cosmic order, so the political order. It is not about taking a philosophical model and reproducing it (e.g. with a monarch representing the Cosmic LSE). Rather, it is about reaching toward an understanding of the cosmos that is there and applying this to the reform of the society we've got. Or as Rudolf Steiner expressed it: "Nothing can bring forth healing for mankind which is not first experienced and thought out in terms of complete reality, and then planted in the social organism."[2]

The modern West thinks government is about politics — the struggle for power between classes, ideologies, parties, and factions; and concerns merely the social institution required to make and enforce laws, manage (in some way) the economy and the environment, and provide public infrastructure and services (the latter including welfare and defense). The first – which is modern politics in advanced countries – is a historically recent phenomenon. With regard to the second, these are necessary functions of government, but government is about so much more than that. In a nutshell, government is about "imitating the One."

The cosmic order as traced in Part I provides for: a) autonomous existence (but with a built-in return magnet), b) growth (to selfhood and then to union with one's author), and c) greater and greater expression of the Good (with each cycle). Government – good government – is about imitating the One in providing for: 1) the basic needs of all citizens, 2) their moral and intellectual growth (the upper range of which is spiritual/transpersonal), and 3) cultural activities, including everyday work, which enable the expression of the Good. As Iamblichus said (he was referring to a wise king – Raj-rishi in Hinduism – but for this we may substitute a wise and good government):

> He guides people more effectively, and even better than that, as true leader, who provides a generous donation of good things and unstinting supply of the means of life and establishes a maximum degree of safety and leisure in living. For this, after all, is the aim of a good ruler, *to cause his subjects to flourish*; and it is precisely then that a leader is distinguished in power above those that he administers, when those who have entrusted themselves to him enjoy a blessed existence. For the common good is not to be separated from the individual good; on the contrary, the individual advantage is subsumed within that of the whole, and the particular is preserved in the universal, in the case of both living things and states and all other natural entities.[3]

And as the Pythagorean Sthenidas the Locrian wrote, seven centuries earlier:

> A king should be a wise man; thus will he be honored in the same manner as the supreme divinity, whose imitator he will be ... a king will imitate the First God in the most excellent manner, if he acquires magnanimity, gravity, and the restriction of his wants to but few things, to his subject exhibiting a paternal disposition. For it is because of this especially that the First God is called the father of both Gods and men, because he is mild to everything that is subject to him, and never ceases to govern with providential regard ... he is nourisher and preceptor of everything beautiful, and the legislator to all things equally.[4]

The Priorities of Government

The ideal and end of human life (not forgetting there is a cosmic context to this) is, in one language, for the "soul" to be in charge. By which is meant the sage-nature in the case of a person, and sages in the case of a society. A sage is a person who has woken up and is full of wisdom – real wisdom, that is, which is knowledge by identity of the extra-philosophical (not just knowledge of the philosophical – an important distinction). A sage is also a person who lives a virtuous life – they don't just preach it (another important distinction). The sage is normally an older person (male or female) who is seasoned in life, but worldliness does *not* equal sagacity.

Elders in the true sense are a rare breed – not that sages can be bred; they are made from within (through contemplative practice), not without. That said, certain things are conducive to the inner growth we are talking about. It is the ultimate purpose of government to ensure that such things are provided as are conducive to that inner growth: the inner growth of society as a whole (which is good for the planet and consistent with our business task in the universe) and of individual members of society (including those serving in government in a political or administrative capacity). Politics, or government in action, is the "art of attending to the soul," as Plato said. As such it is also the "royal art," where "royal" refers to the spiritual monarch (the sage/soul) in every person.

The things that are conducive to "subjects flourishing" are the provision of basic needs, a well-rounded education, and useful cultural activities. The self cannot concentrate on its own inner growth if it is in want of basic needs (which are not limited to sufficient food, clean water, decent sanitation and housing, and health and social care from cradle to grave); if there has not been an education which has led to the production of a person with a moral character, key cultural knowledge, and the ability to think (by which is meant intellection in general, so ideally including the intellectual intuition); and if cultural activities, including everyday work, are not available which are conducive to the end of human life. Right government priorities would then be, and in order:

1. Seeing to it that the basic needs of all people are met (not seeing to it that they *could* be met, but that they *are* met — this is a big difference and concerns the economic system). This also links with Articles 3, 5, 22, and particularly 25(1) of the Universal Declaration of Human Rights (see over the page). This would be the needs of people now and in the future.
2. Seeing to it that there is good universal education, the primary aim of which is *not* the equipping of people with skills and knowledge to service trade and industry (these cultural activities not originally being part of human civilization and having a historical shelf-life — see

next chapter), but the production of people who possess rectitude, an altruistic nature and cosmopolitan identity, are geopolitically and historically informed, and are intellectually developed. This also links with Articles 26(1) and 26(2) of the Universal Declaration.

3. Seeing to it that there is work available which utilizes people's education and natural talents in service not just to their self-interest, but to the wider good, which should be promoted and modeled by cultural and political leaders so that it becomes widely enculturated. This being work which is ultimately the most satisfying and meaningful, so in people's self-interest anyway, with the wider good referring to the collective good of the community and the environment (all other lives on this planet being our kin). This also links with Articles 4 (if we include wage slavery), 23(1), and 27(1) of the Universal Declaration.

Excerpts from the Universal Declaration of Human Rights

Article 3: Everyone has the right to life, liberty and security of person.

Article 5: No one shall be subjected to torture or to cruel, inhuman or degrading treatment or punishment.

Article 22: Everyone, as a member of society, has the right to social security and is entitled to realization, through national effort and international co-operation and in accordance with the organization and resources of each State, of the economic, social and cultural rights indispensable for his dignity and the free development of his personality.

Article 25(1): Everyone has the right to a standard of living adequate for the health and well-being of himself and of his family, including food, clothing, housing, and medical care and necessary social services, and the right to security in the event of unemployment, sickness, disability, widowhood, old age or other lack of livelihood in circumstances beyond his control.

Article 26(1): Everyone has the right to education. Education shall be free, at least in the elementary and fundamental stages. Elementary education shall be compulsory. Technical and professional education shall be made generally available and higher education shall be equally accessible to all on the basis of merit.

Article 26(2): Education shall be directed to the full development of the human personality and to the strengthening of respect for human rights and fundamental freedoms. It shall promote understanding, tolerance and friendship among all nations, racial or religious groups, and shall further the activities of the UN [United Nations] for the maintenance of peace.

Article 4: No one shall be held in slavery or servitude; slavery and the slave trade shall be prohibited in all their forms.

Article 23(1): Everyone has the right to work, to free choice of employment, to just and favourable conditions of work and to protection against unemployment.

Article 27(1): Everyone has the right freely to participate in the cultural life of the community, to enjoy the arts and to share in scientific advancement and its benefits.

The Problems with Existing Government

The three priorities of government are no more than what any right-minded parent would want for their child (for them to have their needs met, a good education, and to go on to be happy in the world, doing good deeds). The general problem with existing government is that it has forgotten its role — which, as Plato said, is to be "shepherd to the human herd." This would be a guardian and directing (steering toward the Good) role which nurtures, protects, and *honors* human members of society. "For everything that is honored flourishes, whereas what is given no honor tends to diminish, and this is the most conspicuous sign of a well-administered regime," wrote Iamblichus.[5] There are also two particular and well-known problems with government today: a) we do not yet have a world government; and b) we have governments more or less serving the wealthy few, on account of generally and naively serving the pursuit of wealth.

With respect to the last, it was not always this way and will not always be this way. Here in this chapter we simply recognize the current state of affairs which, with a roll of his or her eyes, the medieval peasant would also recognize — subjects working to make the nobles richer, so that the nobles can throw the subjects more scraps from their table. Except that the subjects today include governments, and the nobles (who throw governments minimal taxes) include transnational corporations and the super-rich. The poor — the deprived countries of the South, and deprived people in the North — are working harder and longer, and the rich are becoming richer and more powerful. The poor, although they do have more scraps unequally spread between them (such that *extreme* poverty in the South has been reduced in recent decades), are comparatively poorer as the years pass and, the main thing is, still poor. And seemingly powerless too, having no proper guardian in government. As Noam Chomsky observed:

> The large majority of the population, at the lower end of the income/wealth scale, are effectively excluded from the political system, their opinions and attitudes [and general condition] ignored by their formal representatives, while a tiny sector at the top has overwhelming influence [and wealth].[6]

The "tiny sector at the top" could directly refer to the estimated 63 million millionaires (including billionaires) who together own 46% of the world's estimated $464 trillion wealth[7] (a substantial part of which is in offshore tax havens[8]). This compares with the 50% at the bottom who together own less than 1% of it. "Wealth" refers to real assets such as housing, together with financial assets and private pension pots (excluding debts). It's not just about differences in wealth of course; it broadly corresponds also to differences in health, education, security, and life expectancy. This plutocratic capitalist world system (such as we can call it – and a capitalist system always turns plutocratic) peddles two myths: a) that we can all do well out of it, when reality and history beg to differ, and the truth is this system relies on a perpetually needy, discontented, and quarrelsome mass of people, for in the absence of that no money can be made; and b) there is no alternative other than state or syndicalist socialism – yes there is, and it is covered in the next two chapters.

The average person today appreciates the general *wrongness* in this world system. They see governments kowtowing to big business and a general economistic culture which promotes mammonism and dog-eat-dog values – values which they are expected to adopt and even required by their job to practice against their better nature. They see the gross financialization of people and nature – everything being reduced to a monetary value. And they see the harmful effects on society and the environment all of this brings.

Globalization, for many people, and particularly those on lower incomes in the North and in the South, is a nasty word and business. As the former Governor of the Bank of England, Mark Carney, acknowledged: "Globalisation is associated with low wages, insecure employment, stateless corporations and striking inequalities."[9] And as the economist Joseph Stiglitz wrote:

> Now, globalization's discontents in the developing world have been joined by the discontents in the developed world in believing that the system is rigged against them. How can it be that it is rigged against both workers in developing and developed countries? Simple: it is rigged in favor of global corporations. It creates a race toward the bottom among workers, with each country seeking to attract work away from another by offering labor at lower prices. Indeed, standard economic theory explains why economic integration leads to lower wages of less-skilled workers in the advanced countries ... Most of the so-called free trade agreements are not really about free trade. They are instead managed trade agreements — managed for the benefit of corporate interests ... Citizens were told to accept certain changes in the rules of the game because it would make them better off. Now in many countries workers are told they have to accept cutbacks in wages and public services *in order to compete in our globalized world*. The disparity between promises and what has happened has deepened distrust of elites ... and democratic politics.[10]

"Elites" would include economists and intellectuals seen as effete, shallow rooted, and playing the fiddle while Rome burns. The political scientist Patrick Bond also wrote:

> The main beneficiaries of the 'neoliberal' (pro-corporate, anti-social) policies that result from growing financial influence over national states are multinational corporations. Their taxes have been cut and labour costs and environmental regulation lowered by outsourcing or by shifting operations to repressive sites of production. These firms have also moved taxes far beyond state borders, with trillions of dollars' worth of 'illicit financial flows' manoeuvred into offshore financial centres, leaving governments with rising budget deficits and their social sectors experiencing permanent cost-cutting pressures.[11]

As a result of such neoliberal policies and antisocial behaviors, governments have less money to spend on public services (including welfare and overseas development aid), and the solution proposed for this service deterioration is privatization — being taking over by the same multinational corporations who: a) know governments have less money and insufficient global cooperation to hold them to account, b) prefer and preach self-regulation anyway (which right-wing governments go along with as per their ideology), c) demand tax-friendly policies on the argued basis that this spurs economic growth (which the populace is supposed to think benefits everyone and is the purpose of human life — both are untrue), and d) seek compensation from governments (written in to the terms of trade deals) for changes in government policy that might enhance public wellbeing but harm private profits. One reaction to this sorry state of affairs is the global justice movement, and in support of this the "Global Democracy Manifesto" is included over the page.

One might think that the Left would be in the ascendancy as a result, but apart from some resurgence in Marxist thought and a well-intentioned but not-quite-the-answer ecosocialism, together with a majority of young people and the more educated who are either by instinct or schooling cosmopolitan and egalitarian, we have seen the deprived and the disaffected (and the less educated) turning especially to the populist Right, and being less internationalist and liberally-minded — prey to demagogues sowing fear and division. Hence such things as the 2016 Brexit vote in the UK and the election of Donald Trump in the US. What the deprived and disaffected want of course, and are not getting, is that proper guardian in government. To quote from the Global Solutions report to the 2017 G20 summit:

> Driven by the interlocking forces of globalization, technological advance and financialization, economic success is no longer in lockstep with social success. Evidence abounds. In many advanced industrialized countries, the growth in aggregate real income has been accompanied by rising inequalities and stagnant living standards for the common folk. This ... helped generate the discontent that influenced the recent US election outcome and the UK's decision to leave the EU.[12]

If only the "common folk" better understood that further liberalization is not at all in their interest. If only the common folk better understood that the tenure of the super-rich and the transnational corporations is further sustained by these "nobles": a) encouraging the poor to blame each other for their condition — whether in the form of working-class immigrants and asylum seekers, overseas aid recipients, or welfare claimants, including the disabled; and b) promoting nativism and anti-supranationalism — for transnational governments would have more power over transnational corporations and be better able to collect taxes from them and the super-rich, so for these "nobles" supranationalism is the enemy. They rule by dividing nations and communities — the oldest trick in the book.

It is also the case that many so-called social democrats have been just as enamored of Big Money as the Right, and have adopted the same disdainful attitude toward the welfare state — or they have simply been unable to articulate much of an alternative. This is because they have fallen into the trap of accepting that economic growth is the goal *when it is not* (see Chapter 3). Both public reactions — the global justice movement, and the turning to populist nationalism — are understandable, and neither is wholly effective in changing the status quo. However, as the former grows and intensifies and eclipses the latter — as it will — so there will be an erosion breakthrough. It is realistic to expect this will take decades: Rome wasn't built in a day. We are advanced in co-perceiving the problem. We are not so advanced in co-perceiving the deeper causes and the general solutions to the political crisis.

The Global Democracy Manifesto

> Politics lags behind the facts. We live in an era of deep technological and economic change that has not been matched by a similar development of public institutions responsible for its regulation. The economy has been globalised but political institutions and democracy have not kept pace. In spite of their many peculiarities, differences and limitations, the protests that are growing all over the world show an increasing discontent with the decision-making system, the existing forms of political representation and their lack of capacity for defending common goods. They express a demand for more and better democracy.
>
> Global welfare and security are under threat. The national and international order that emerged from the end of World War II and the fall of the Berlin Wall has not been able to manage the great advances in technology and

productive systems for the benefit of all humanity. On the contrary, we are witnessing the emergence of regressive and destructive processes resulting from the economic and financial crisis, increased social inequalities, climate change and nuclear proliferation. These phenomena have already affected negatively the lives of billions of human beings, and their continuity and mutual reinforcement menace the peace of the world and threaten the survival of human civilisation.

Global crises require global solutions. Within a social universe determined by globalisation, the democratic capabilities of nation-states and international institutions are increasingly restricted by the development of powerful global processes, organisations and systems whose nature is not democratic. In recent years, the main national and international leaders of the world have been running behind global events. Their repeated failures show that occasional summits, intergovernmental treaties, international cooperation, the multilateral system and all the existing forms of global governance are insufficient. The globalisation of finance, production chains and communication systems, and the planetary power reached by destructive technologies, require the globalisation of the political institutions responsible for their regulation and control, and the global crises require coherent and effective global solutions. That's why we call for the urgent creation of new global agencies specialised in sustainable, fair and stable development, disarmament and environmental protection, and the rapid implementation of forms of democratic global governance on all the issues that current intergovernmental summits are evidently incapable of solving.

We need to move forward to new, more extensive and deeper forms of democracy. The current model of technological-economic globalisation must give way to a new one which puts these processes at the service of a fairer, more peaceful and more humane world. We need a new paradigm of development which has to be sustainable on a global basis and which benefits the poorest of humanity. In order to avoid the deepening of global crises and to find viable solutions to the challenges posed by globalisation we must move forward to more extensive and deeper forms of democracy. The existing national-state organisations have to be part of a wider and much better coordinated structure, which involves democratic regional institutions on all the continents, the reform of the International Court of Justice, a fairer and more balanced International Criminal Court and a United Nations Parliamentary Assembly as the embryo of a future World Parliament. Yet, this institutional change will not be successful if it only accrues from the actions of a self-appointed elite. On the contrary, it must come from a socio-political process open to all human beings, with the goal of creating a participative global democracy.

Globalising democracy is the only way to democratise globalisation. Beyond our differences about the contents and appropriate methods to move towards a fairer and more stable world order, we the signatories share a strong commitment to the development of a global democracy. On behalf of Peace, Justice and Human Rights we do not want to be governed at the world level by those who have only been elected to do so at the national one, neither do we wish to be governed by international organisations which do not represent us adequately. That is why we work for the development of supranational political spaces and for regional, international and global institutions that live up to the challenges of the 21st century; institutions that express the different viewpoints and defend the common interests of the 7 billion people who shape humankind today.

We ask every human being to participate in the constitution of a global democracy. We share the appeal to "unite for global change" and for "real democracy" with the world social movements. Both postulates express the growing rejection of being governed by political and economic powers on which we have no influence. Autonomy and self-determination are not only valid at the local and national level. That's why we champion the principle of the right to participate in the making of fundamental global decisions that directly affect our lives. We want to be citizens of the world and not its mere inhabitants. Therefore we demand not just a local and national democracy, but also a global democracy, and we commit to work for its development and call on all the political, intellectual and civil-society leaders of the world, all the democratic organisations, parties and movements, and all persons of democratic persuasion on the planet to actively participate in its constitution.[13]

Spiritual Aristocracy

Both these words come with a lot of baggage so we will tread carefully. The ideal form of government is a no-brainer – and we all know this because it is also the ideal form of government in ourselves: one led by the wise and virtuous (our highest selves). But sages are made from within, not without, so the principle of spiritual aristocracy is *not* a call for artificially setting up some philosophocracy (and certainly not an *ecclesiocracy* as sages have no religion – this is where the Islamist errs, for one), but mutually recognizing the *to-be* rule of the qualitatively superior and *reaching toward that* (basically, going with the return magnet). Plato understood that a spiritual aristocracy (where this refers to a team of the wise and not to a group made up of a certain class) is the ideal form of government, but an even higher form of government is the rule of the Good (where that has become "automatic" in the person due to his or her ultimate self-development). Plato took a dim view of democracy, and for this reason is sometimes misread as an authoritarian or oligarchist, but that would be democracy alone or unimproved.

Fools electing fools to rule over them (and the numerically fewer non-fools) is ever the concern with democracy alone. The risk of ochlocracy (or even kakistocracy) is greatly reduced where: a) a society, through its worldview, is more qualitatively than quantitatively oriented, and thus people are more inclined to reach toward the qualitatively superior; and b) a state provides for a good universal education (including higher education which should equally be free) such that, to put it bluntly, there are fewer fools. Together with necessary institutional checks and balances, and constitutional safeguards (such as separation of powers and time-limited office), adding the principle of meritocracy (or professional competency) to improve a democratic system is uncontroversial — all sensible people recognize the requirement of a certain level of professional competence as well as moral uprightness in their leaders and officials. The principle of spiritual aristocracy is *not* incompatible with a democratic system, for the two are referring to different things: the first refers to the to-be rule of the wise and the process of reaching toward that (applying to voters and the elected alike); the second refers to the to-be election of rulers (and removal if they're no good) by the people. Ordinary democracy, plus safeguards, plus meritocracy, plus spiritual aristocracy (as defined here) — this equals an improved democratic system.

> In the intelligible place, the idea of the good is the last object of vision ... and this must be beheld by him who is to act wisely, either privately or in public.
>
> — Plato[14]

Reaching toward the ruling of the wise implicates the philosophical practice that leads to the "last object of vision" in Plato's quote above. The Idea of the Good is the collective term for the universals manifested on all the levels of being. The Idea of the Good includes the Idea of Love. Aldous Huxley wrote in 1944: "Our present economic, social and international arrangements are based, in large measure, upon organised lovelessness."[15] The best we can say, eight decades on, is that progress has been patchy – with a collective "outsourcing," it would seem, of our conscience to the United Nations, which at the same time we have made sure is underfunded and underpowered so that we can complain about its ineffectiveness. The Idea of the Good also includes the Idea of Beauty. Echoing the point that it's not about completely ripping up what we've got and starting over, but rather reforming and conforming what we've got to wiser principles, Manly Hall wrote:

> It is not necessary that we tear down the entire structure of our present system or revert to some savage type and start anew. It is merely necessary that we tincture utility with beauty; that we add the soul qualities of symmetry and grace to the products of our schemings.[16]

World Government

> For nature has given us no country, as it has given us no house or field ... but every one of them is always made or rather called such a man's by his dwelling in it or making

> use of it ... But Socrates expressed it better, when he said, he was not an Athenian or a Greek, but a citizen of the world.
>
> — Plutarch[17]

Our "schemings" today include progressing the idea of world government by building upon the institution of the United Nations. The general idea of a world government is as old as the hills and is supported by many religions, including Catholicism and Bahaism. The idea has been supported by many twentieth-century luminaries, including Bertrand Russell, Albert Camus, Albert Einstein, Mahatma Gandhi, and Martin Luther King, in a line that can be traced back through earlier figures, such as Anarcharsis Cloots, Francisco de Vitoria, and Philo of Alexandria. The most recent and developed form of the idea, further to two earlier proposals of note (the Preliminary Draft of a World Constitution[18] and the Constitution for the Federation of Earth[19]), is the United Nations Parliamentary Assembly proposal. Recognizing that presently the bodies of the United Nations and international organizations are occupied by state-appointed officials — not by elected representatives — the proposal is:

> A United Nations Parliamentary Assembly (UNPA) for the first time would give *popularly elected representatives* a formal role in global affairs. As an additional body, the assembly will directly represent the world's citizens and not governments. Initially, states could choose whether their UNPA members would come from national parliaments, reflecting their political spectrum and gender equality, or whether they would be directly elected. Eventually, the goal is to have *all members directly elected*. Starting as a largely consultative body, the rights and powers of the UNPA could be expanded over time

> as its democratic legitimacy increases. The assembly will act as an *independent watchdog* in the UN system and as a democratic reflection of the diversity of world public opinion. In the long run, once its members are all democratically elected, the assembly could be developed into a *world parliament* which — under certain conditions and in conjunction with the UN General Assembly — may be able to adopt universally binding regulations. In short, the UN should evolve from what many believe to be a generally ineffectual "talk-shop" into a viable democratic and legislative body.[20]

The eventual world parliament would then be an alternative to US hegemony or to a multipolar world order, and end the outdated Security Council arrangement whereby just five countries (the US, Russia, China, the UK, and France) each hold a power of veto on any change to the UN Charter and thus effectively to global governance. We might reflect, though, that *simply* having a global authority would not be enough. Apart from the challenge of just relying, if we did, on democracy unimproved, we would need a better understanding of the role of government. One might use a cooking metaphor – the soul is the natural sweetness in the dish (society) brought out by the right preparation of the ingredients (citizens). Good government is a good cook. We would want a world government that understands this. Guiding ideas are already present in the Preamble to the UN Charter, the Preliminary Draft of a World Constitution, the Universal Declaration of Human Rights, the Preamble to the Earth Charter, and of course the teachings of the world's religions.

One of the main objections to the idea of a world government is that there would be no escape if it became a totalitarian state. But then we have a totalitarian, corporatocratic state today, many would say. In any event: a) only a global government can manage a global economy — and the absence of that (a globally managed economy) is our biggest problem, and b) all the more reason why a world government would then need to "get off on the right foot" when it comes to understanding its role and basing itself on the right principles (likely encapsulated in a World Constitution). A totalitarian state has a "compulsory" ideology, whereas the government we are envisaging merely promotes human development beyond all ideologies — bringing out the soul in citizens, not brainwashing and controlling them.

Another objection has in mind the sheer political difficulties involved in, and the likely nationalistic flare-ups caused by, the very attempt. On this objection, the author does not believe humankind is incapable of overcoming these difficulties and managing these flare-ups, nor that it is unwilling to make the attempt. A 2023 poll showed 60% support for a global parliament,[21] no doubt linked to what most cultural observers agree there is today — the emerging of a world *society* (if not yet a world *demos*), and a movement toward world *domestic* laws (not just *international* laws). The author believes that suitably planned educational and change-management activities could allay concerns, reduce resistance, and mitigate flare-ups. We have, after all, done this sort of thing before with the creation of nation-states in the first place, and then of supranational organizations such as the EU.

We would need to advance the benefits of a world government and cultivate a planetarist mindset, building upon the existing recognition of the environmental crisis. This may seem far away from a world today where around a third of people live under authoritarian regimes, but then we are talking about the world's citizens en masse and intercommunicating as never before — including on the wrongness of the current world system — and history shows that democratic revolutions can happen relatively quickly and painlessly. Also, we ought to recognize that the long-term trend toward world democratization does continue apace. It is like the difference between the weather and climate change. There is the day-to-day mixed bag of weather events and conditions, just as there is the day-to-day mixed bag of political events and conditions. Distinct from these mixed bags are the long-term trends. One is heartened by a previous poll showing a majority of people seeing themselves more as global citizens than as citizens of their own countries.[22] Skeptics and pessimists, take note!

Both nationalism and the Westphalian nation-state are coeval and interlinked psychological and political constructions. Constructions of the adolescent and adult minds, that would be. We borrow here from the psychologist Robert Assagioli, where there is: 1) *personal* development, which is to do with outing a mature adult self — a self that has left behind, or is in control of, the "child mind" (which is susceptible to such diseases as phobias, obsessions, and compulsive urges) and the "adolescent mind" (which is susceptible to such diseases as bigotry, racism, and fundamentalism). And then there is: 2) *transpersonal* development, which is to do with outing the fully developed human being — which is the sage. This is the person who has left behind, or is in control of, the adult mind, which is yet susceptible to such diseases as egotism (and therefore selfish political ambition), moral cowardice (leading to the corruptibility of the politician), and rationalism (being led by the rational mind, rather than or more than the intellectual intuition). Nationalism fits somewhere between the diseases of the adolescent and the adult minds and is "treatable." We know this because we can observe the general success of the EU project.

Another objection to a world government is that it would be overly bureaucratic and remote from local life. This is an objection partly based on the liberalist view — a view which is widely held and laudable in its commitment to freedom and equality, but at the same time is almost the opposite of ancient wisdom in its recognition of "mundane man" only — a creature which is "naturally" selfish, greedy, competitive, and parochial, as in the Hobbesian depiction of the human being. What government is about in this view is going with the flow of *this* nature — with nation-states likewise modeled on this — rather than going with the flow of nature as ancient philosophers such as Plato and Pythagoras understood it.

Liberty, in the liberalist (and neoliberalist) view, is freedom to be oneself, as individual or nation-state, as an essentially *ignoble* creature (ignoble in its pursuit of wealth and power, that is) within an equally "natural" society built around the marketplace and market values. Government in this "ideal" society performs functions required to keep the capitalist show on the road — but that's about it. This contrasts with the other view of liberty as the freedom to be oneself *as a noble soul,* within a community of fellow noble souls: this is the proper ideal society, which is to say, *the* ideal society. As Plato wrote:

> We must say then that this end of politic action [the ideal society] is then tightly woven, when the royal art, connecting the manners of brave and temperate men [and women] by concord and friendship, collects together their life in common, producing the most magnificent and excellent of all webs.[23]

The localist and syndicalist concern for community-level sustainability and relative autarky is a genuine one. However, in such a subsidiarity model as envisaged below, there is no reason why the general principles of communalism and self-government could not coexist with higher levels of government where there is a common planetarist mindset. And where also — which is to say as long as — the whole planetary government system manifests "justice" as the ancient philosophers understood it (which includes social justice). Getting it right at the top, with a world government and its Constitution — and the general orientation of its parliamentarians and the laws they pass to the Good — is key.

Another genuine concern is the loss of cultural diversity, but here we must be careful. The outcome sought is not plurality of cultural expressions per se, but plurality of cultural expressions of an acceptable kind. The one thing we all share in common is the human experience, and the one thing we all know and can agree on is that all other things *by comparison* (such as gender, race, religion, and nationality) are secondary to this. The evidence of us knowing and being able to agree about this is found in such developments as the abolition of slavery, and the existence of the "international community" today that reacts largely with one mind to terrorist atrocities, natural disasters, illegal invasions, and war crimes wherever they occur. A cultural homogenization is already taking place across local and regional spheres with economic globalization — but the point here is that cultural diversity and a planetary monoculture are *not* mutually exclusive. It all depends on what the latter is based on.

If it is based on the one thing we all share in common (or the two things if we include our planet), then here is a starting point for a higher level of planetary culture not replacing but *conditioning* the lower level (as a responsible and thoughtful adult conditions the irresponsible and thoughtless teenage nature within him/herself). We might reflect here that a more enlightened civilization, such as the one we can imagine the extraterrestrial coming from (and how is it we can easily imagine this but at the same time think it unattainable for us?), would not have retained in its diversity of cultural expressions any that mistreat people or animals (or the equivalent on its planet). Freedom of expression and the freedom to live is one thing. The freedom to hurt and the freedom to not let others live is unacceptable.

Holarchy and Subsidiarity

The universe is holarchically structured. Being is organized into successive ranks, with each level being subordinate to the one above. Not just a hierarchy but a holarchy – the ranks featuring holons, which are simultaneously wholes and parts, like cells within organs within bodies, or towns within counties within provinces, with the good of one holon affecting the good of another, above or below. This structure may be imitated somewhat with a world government at the top and, beneath that, regional and national governments, and, beneath those, provincial and local governments. In the vertical universe, which is to say in wider nature, the lesser is being ideated or consciously experienced by the greater as its vehicle of expression – whether we have in mind, as examples of this, a person and his/her body, or the Earth LSE and the planet of human, animal, plant, and mineral kingdoms.

Holarchy and subsidiarity are familial, but they are not the same. In the first, the lesser is being ideated and brought into existence by the greater; the lesser is living itself but, at the same time, since it is that in which the greater resides and acts (as we do our corporeal bodies and an author does his or her imagined characters), so the greater lives it. In the second, the lesser is not brought into existence by, nor is it in any sense lived by, the greater. We have subsidiarity in the European Union, but we cannot say the EU lives France as an author lives his character. Rather, with subsidiarity, the lesser confers authority over itself (on certain matters) to the greater – matters which are accepted as being best dealt with at a higher level. With member states of the EU, there is no requirement that these are matters that *can* only be dealt with at the European level. At the same time, the principle "works the other way" in accepting that matters best dealt with at a state level are dealt with at that level, without the requirement that all matters are dealt with at the state level unless they cannot be.

One can imagine (and the United Nations Parliamentary Assembly proposal does) the same principle operating with respect to a world government — nations worldwide conferring authority over themselves to the greater body on matters accepted as being best dealt with at a global level (such as tackling climate change, protecting the environment, and ensuring human rights), without the requirement that they be matters that *can* only be dealt with at a global level. Form follows mindset, though, so mindset is where we need to focus our efforts. We have spoken of the need for a planetarist mindset, and it is true that "think global, act local" is already in the public lexicon. But the "global" needs to be better understood as referring more broadly to the universal good — which ultimately points to the Idea of the Good. This goes beyond the usual political and environmental management thinking, because it is philosophical and because it does reach vertically. It goes beyond the materialist worldview, in other words, which is why it is The Big Challenge for today's (largely scientific-atheist) global thinkers.

We would ideally want holarchical thinking in politicians' minds. We would want the national politician thinking simultaneously and in a far-sighted way of the regional and global communities "upstream," and provincial and local communities "downstream," as well as thinking of his or her nation. And we would want the provincial politician thinking of the national, regional, and global communities upstream, and the local communities downstream, as well as thinking of his or her province. In this way, parochialism is staved off, and voters/citizens can join politicians doing the same. This is consistent with the sage who sits at the center of the circle with the cross in it, looking upwards at the cosmos and the

Aevertinal, downwards at corporeal nature and ordinary time, and outwards at society and the personal/transpersonal journey which takes place in Consciousness-time. This journey is further picked up in the next chapter, where we will consider the past and the future of civilization from an esoteric perspective.

Chapter 5

Economics, History, Society

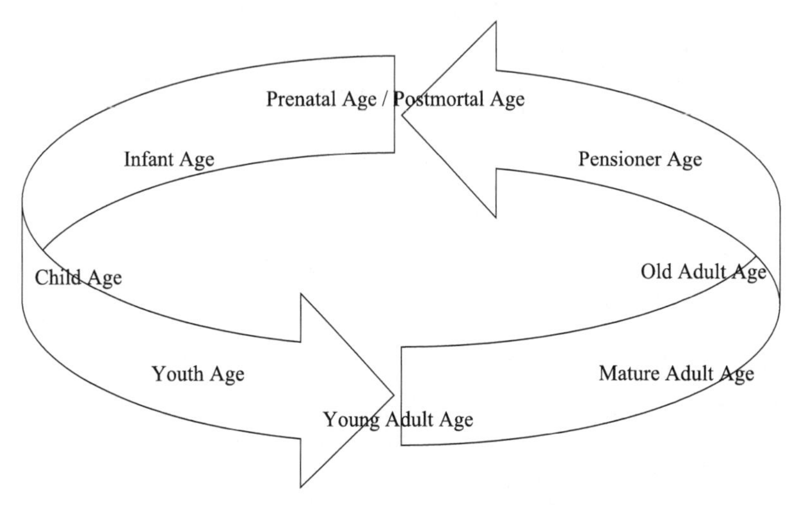

Figure 2: The Ages of Humankind

From prenatal age to young adult age: the outward arc that brings Jack into being as the objectively real person he thinks he is. From young adult age to postmortal age: the inward arc that brings Jack out of being himself into self-conscious union with his author. Between the prenatal and postmortal ages lie the "seven stages of life." The outward arc traces the growth of an increasingly personal and centralized sense of self, and thus personal possessiveness, manifesting collectively as an increasingly competitive/non-sharing society. The inward arc traces the growth of an increasingly *trans*personal and *de*centralized sense of self, and thus personal *un*possessiveness, manifesting collectively as an increasingly cooperative/sharing society.

Introduction

Government is about "imitating the One" in instituting a society that provides for the basic needs of its members; their moral and intellectual growth; and cultural activities, including everyday work, that are geared toward and enable the creation and maintenance of a beautiful, just, and peaceful civilization. The first part concerns the economic system. Economics is about "household management," as per the Greek word *oikonomia*. It is about the management of resources within the household, ensuring the basic needs of everyone living in it are met. Our "household" is the human world with its 8 billion "family" members. Currently, 1 in 8 of these struggle with hunger even though we have more than enough food (supplies of necessities in general) for all members. Clearly we are doing a bad job of household management.

One can imagine a family shipwrecked on a desert island. What would the parents think and accept their *governing and shepherding* role to be? Answer: to manage the natural resources of the island, ensuring a sustainable and non-polluted supply of them, and managing their distribution and use within the household so that none will go without and all can flourish (the children thus growing up to add to the general wellbeing of the family and the island). The situation we have today resembles the parents giving over their role to the children (private economic actors), who impudently and imprudently compete for ownership of the resources, manage them only for profit, and have little or no regard for tomorrow and whether all will "go with" (i.e. have enough) and flourish.

Metahistorically (or psycho-historically), our present economic system belongs to the young adult age and its type of consciousness. Previously there were four other economic systems belonging to the prenatal, infant, child, and youth ages. And ahead there are four economic systems belonging to the mature adult, old adult, pensioner, and postmortal ages. We are not tracing — with respect to the previous ages — ordinary linear history so much as successive types of consciousness that cut across exoteric history and which can be appreciated as existing within ourselves (and observed when we look at the behavior of our young). The Hinduist philosopher would likely recognize in the infant, child, youth, and young adult ages, the Krita, Treta, Dwapura, and Kali yugas. This ambit may also be compared with Hegel's historicism where: "The life of the ever-present Spirit is a circle of progressive embodiments, which looked at in one respect still exist beside each other, and only as looked at from another point of view appear as past."[1]

With respect to the future ages, we have already begun the mature adult age — it's just that there's overlap with the end of the young adult age. *This explains the general crisis of today.* The terms "mature adult," "old adult," and "pensioner" are the author's invention and used to signify the general idea of growing older and wiser. They are not meant to suggest, for it would not be the case, that a septuagenarian today necessarily has a different type of consciousness from a tricenarian today. We are merely tracing further types of consciousness. Because there is a connection between the ages facing each other in the diagram (and a general logic to the whole), some features of the future as well as the past can be sketched. Every type of consciousness produces new results. Collectively, *it compels a new sociopolitical order,* including a new economic system. This is the opposite of the Marxian view which has it the other way around (i.e. in the Marxian view socio-economic conditions determine human consciousness).

The Prenatal and Postmortal Ages

In the prenatal age, there is consciousness but not *self*-consciousness. Self-consciousness is required for any sense of possessiveness (there has to be a sense of "me" before there can be a sense of "mine"), and here there isn't any. Self-consciousness is required for any sense of exchange with another (it is not required for any exchange with another, but it is required for any *sense* of exchange with another). What holds for the microcosm (the individual) also holds for the macrocosm (the society). There is no sense of "us" and therefore "ours" in this earliest Age of Humankind — there is what we might call unselfconscious collective unpossessiveness in the "herd," and there is unselfconscious automatic exchange between members of the society. This is the economic system, if we can call it such, of this age.

The difference between this age and the postmortal age at the end of the cycle is that in the latter there is self-consciousness, but not as the character-self (because Jack has fully woken up). In a society where there is a shared self-consciousness as the author, there can't be any sense of *personal* possessiveness (for although there is self-consciousness, it is not as separate personal selves — and there can only be a personal sense of possessiveness where there is a personal sense of self). In the postmortal society, there is *self-conscious* collective unpossessiveness and *self-conscious* automatic exchange between members of the society.

In the prenatal age, there is *un*selfconscious union with the divine. By "the divine" is meant the Earth LSE, and this relates to the *henosis* experience. But here's the thing: that experience is of a greater entity looking out from behind the person's eyes at Itself, with the person an intelligent and fully self-conscious participant in that experience. In the prenatal age, there is no intelligent and fully self-conscious person present — or any self-conscious person present at all. There is "divine union," but without the person knowing/appreciating/understanding it.

There is a complete and automatic free sharing in the prenatal society, without any sense either that "I am" or "we are" sharing. This is ultimately where we are trying to get back to — an original paradise indeed. But returning to it at the end of the cycle would now be done in a fully intelligent and self-conscious way, such that: a) we know we are sharing, but b) we still do so in the same automatic way, because c) to recall what was said in the Introduction to this book, our mutual self-sense is not now as essentially separate selves but *as* coincident in essential identity. We thus have, in the postmortal age, the ultimate economic system of self-conscious collective unpossessiveness and self-conscious automatic exchange. Beyond the postmortal age — outside the cycle — there is simply the life of the author.

The Infant and Pensioner Ages

> Happy the age, happy the time, to which the ancients gave the name of golden, not because in that fortunate age the gold so coveted in this our iron one was gained without toil, but because they that lived in it knew not the two words "mine" and "thine"!
>
> — Miguel de Cervantes, from *Don Quixote*

In the next age — the infant age (also known to the ancient philosophers as the "Golden Age") — there is the *beginning* of self-consciousness. There was no self-sense in the prenatal age. Here in the infant age there is the beginning of a *physical* self-sense. This allows for the beginning of a sense of "me" and therefore also "mine," and as a society, a sense of "us" and therefore also "ours." We are now talking about basic human societies — the first that can properly be called such, for they now have some sense of themselves *as* societies. There is no social division or stratification yet, and only a sense of territorialism ("our land," "our resources"). This sense is not strong enough to allow for sedentism and farming/producing from that territory. Here in the infant age, individual members of societies can and do have the beginning of a sense of private property ("my blade," "my fur"), but this is not strong enough to allow for taking *responsibility* for such items, and there is certainly no concept of *owning* them. As it is with the individual member of society, so it is collectively for the society.

Economic exchange between individual members of society can and does begin in the infant age, but the economic system is what we might call a "free gift economy." There is no return-based exchange of any kind yet. This would require a firmer sense of "me," and correspondingly "you" and "it" (the thing exchanged). There is not yet, in the mind's eye of the individual, a clear existential division between giver, givee, and the thing given. This is the age we see with very young children who freely give to their peers a toy or a sweet. There is no thought of what they may get back from this act. There's no clear sense of "I'm giving you this thing." It's a "magical" exchange of an indefinable thing between indeterminate parties.

The infant age contrasts with the pensioner age in this way: in the former, there is the *beginning* of self-consciousness, whereas in the latter there is the *ending* of self-consciousness — self-consciousness as an essentially separate personal self, that is (Jack has almost fully woken up). In the former, this allows for the *beginning* of trade and industry (economic exchange of a magical kind, and "industry" of a hunter-gatherer kind); in the latter, this allows for the *ending* of trade and industry — a "retirement" from this cultural activity. In the former, this allows for a free gift economy based on an *unknown* (but instinctive) interconnectedness; in the latter, this allows for a free gift economy based on a *known* interconnectedness. The "new golden age" that is the pensioner age equates with Plato's "end of politic action." This is the ideal society of men and women connected by the "royal art," living their life in common and producing the "most magnificent and excellent of all webs." Ovid described the original Golden Age thus:

> The golden age was first; when Man yet new,
> No rule but uncorrupted reason knew:

And, with a native bent, did good pursue.
Unforc'd by punishment, un-aw'd by fear,
His words were simple, and his soul sincere;
Needless was written law, where none opprest:
The law of Man was written in his breast:
No suppliant crowds before the judge appear'd,
No court erected yet, nor cause was heard:
But all was safe, for conscience was their guard.
The mountain-trees in distant prospect please,
E're yet the pine descended to the seas:
E're sails were spread, new oceans to explore:
And happy mortals, unconcern'd for more,
Confin'd their wishes to their native shore.
No walls were yet; nor fence, nor mote, nor mound,
Nor drum was heard, nor trumpet's angry sound:
Nor swords were forg'd; but void of care and crime,
The soft creation slept away their time.
The teeming Earth, yet guiltless of the plough,
And unprovok'd, did fruitful stores allow:
Content with food, which Nature freely bred,
On wildings and on strawberries they fed;
Cornels and bramble-berries gave the rest,
And falling acorns furnish'd out a feast.
The flow'rs unsown, in fields and meadows reign'd:
And Western winds immortal spring maintain'd.
In following years, the bearded corn ensu'd
From Earth unask'd, nor was that Earth renew'd.
From veins of vallies, milk and nectar broke;
And honey sweating through the pores of oak.[2]

Not merely the infant-age type of consciousness but also the infant-age type of economic system (or gift economy generally – with no or only a little thought of return) continues into our modern age in exchange between loved ones, and in exchange between members of remaining hunter-gatherer societies. As the anthropologist Marshall Sahlins wrote: "The gift is the primitive way of achieving the peace that in civil society is secured by the State."[3] The gift economy underpins *all* societies; we only need to think of the vast amount of unpaid parenting, caring, and volunteering work that is not counted in gross domestic product (GDP) but makes up a vast amount of productive activity. The economist John Gowdy also adds, as a reminder that the capitalist system we currently have is not the system we've always had, nor is it the system that *is* human nature but merely is *in* human nature (along with other systems):

> The more we learn about hunter-gatherers, the more we realize that the cultural beliefs surrounding modern market capitalism do not reflect universal "human nature." Assumptions about human behavior that members of market societies believe to be universal, that humans are naturally competitive and acquisitive, and that social stratification is natural, do not apply to many hunter-gatherer peoples. The dominant school of economic theory in the industrialized world, neoclassical economics, holds these attributes to be essential for economic advancement and affluence. It is true that hunter-gatherer societies show a wide variety of patterns of culture, some less egalitarian and some less "affluent" ... Yet the very existence of societies living adequately, even happily, with no industry, no agriculture, and few material possessions offers a challenge to the concept of human nature held by most economists.[4]

The Child Age

In the child age (also known to the ancient philosophers as the "Silver Age"), there is now a *developed* physical self-sense (the birth of the "I") and the beginning of an emotional self-sense. Having a definite sense of a physical "me" allows for an exclusive sense of "mine"; and collectively, a definite sense of a physical "us" allows for an exclusive sense of "ours" (referring to the society's territory and what's in it). This enables property to be cared for (including the individual's body), and the sense is strong enough to allow for farming/production from the land (this is the horticultural/pastoral/agricultural revolution, metahistorically speaking). It is production for need only, though.

In the child age, individual members of societies have a sense of private property strong enough to take responsibility for their property. As it is with the individual member of society, so it is collectively for the society. This allows for the invention of *government* (the agent of the people – a ruler or rulers – carrying out that responsibility). There is still no concept of ownership, though, so between individual members of society it is a case of "what is mine is also yours should you need it," and within society and managed by the ruler(s) there is a pooling and sharing of resources as required. As there is not yet a *developed* emotional self, there is not yet an individual that can be sufficiently self-oriented (or to put it the other way, insufficiently group-oriented). This group consciousness does not come from group culture, but rather the other way around.

The child age is the age of nature religion, where the world (which is now experienced as less amorphous and more external than it was in the infant age) is the embodiment of the divine. There is a "symphony" between: a) the land being the people's, and b) the people being the land's. It is this way because the world is divine, and so with respect to the first we have the divine serving the people, and with respect to the second we also have the divine serving the land. Rulers in the child age held the sacred trust to be guardians of the land, the people, and the overall "religious" culture (not that there was any sense within the community that this was being religious).

Non-free gift exchange can and does begin in this age; it is like what we see with somewhat older children who still give the toy or sweet but have some thought of *reciprocation obligation*. The child age society is a non-growth, subsistence-based society. In it, members live in houses and have jobs, unlike in the infant age, but these land-based and craft-based vocations are fully rewarding and meaningful in: a) their place in the community, and b) that community living in a sacred world where everything has spiritual meaning. The Silver Age is the Arcadian rural idyll. Not merely the child-age type of consciousness but also the child-age type of economic system continues into our modern age in some rural communities — in monasteries and ecovillages for instance. Behind every "back to the land" dream and agrarianist philosophy is the child age.

The Youth Age

In the youth age (also known to ancient philosophers as the "Bronze Age" — but not one and the same as the Bronze Age of exoteric history), as well as a developed physical self-sense there is now a developed emotional self-sense (and the beginning of an intellectual self-sense). A developed emotional self-sense means the self can now be sufficiently self-oriented — and that means it can be self-centered, self-indulgent, self-acquisitive. From a certain perspective, human civilization goes downhill from this point. For together with the already exclusive sense of "mine," and in the absence yet of much of a thinking nature that can temper the emotional nature, this is the age where there can be — and was — something new in societies: pleasure, power, and wealth-seeking *personalities*.

Pleasures of the carnal type, that would be (this being the youth age after all), and power *over* others (both within the community and with respect to conquering other communities).

And wealth-seeking well beyond material need. There was no material concept of wealth before – there is now. Private ownership makes its violent, if natural, appearance in this age, both as a concept and as an action taken, and autocratic rulers may now serve themselves rather than the people (with the former sacred trust replaced with a "divine right" to rule over the land and the people). This is the age of increasing social stratification and organized (though not yet doctrinally codified) religion. This is the age of militarism, of proto-states acting like street gangs, and "might is right." The youth age society was a pecking-order society, like a prison or a schoolyard today.

With respect to the religiosity of the youth age, the world is still *given* by the divine, but nature in general is no longer necessarily the embodiment of it. This new theistic consciousness, coupled with the new selfish consciousness, meant there could be: a) expanded production of the land which now perhaps was just that (a source of wealth not deserving of reverence); b) the new societal features of inequity, poverty, slavery, and pharaonic and feudalist masters; and c) the appearance of a new type of economic system – exchange for personal wants and, increasingly as the *folkland* was done away with, needs (with a form of money developed as the medium). In the youth age, material wealth was "reallocated" from society to individuals – just as self-identity was. In the mature adult age, material wealth is reallocated back from individuals to society – just as self-identity (while retaining individual identity) is.

The Young Adult Age

> And deceived by the false science of reasons, they will direct their hearts towards everything mean and low … And no one will, at that time, be a giver (of wealth or anything else) in respect to anyone else … And the inhabited regions of the earth will be afflicted with dearth

> and famine, and the highways will be filled with lustful men and women of evil repute ... And when the end of the [Kali] *Yuga* comes, urged by their very dispositions, men will act cruelly, and speak ill of one another ... And people will, without compunction, destroy trees and gardens ... And men will be filled with anxiety as regards the means of living ... And the low will become the high, and the course of things will look contrary.
>
> — From the *Mahabharata*

The next age in the sequence is the one we are currently at the very end of — the young adult age, aka the "Iron Age" in ancient philosophy (and identifiable with the Kali Yuga in the above quote). In the young adult age, as well as having a developed physical self-sense, and a developed emotional self-sense, there is now a developed intellectual self-sense (and the beginning of a *trans*physical self-sense — more on this in the next section). We have now reached the midpoint of the cycle, with Jack-the-character the objectively real person he thinks he is. He thus proceeds, over a linear time he newly conceives and measures, to fully explore, map, study, and exploit his "objectively real" world.

The young adult age is the "dark age" in that the character is now furthest away from the "light" of conscious union with his author. It marks the end of the psychological Fall of Man. The young adult age is the age of *Homo economicus* citizens. We can say *citizens* now, because this is the age where there can be *states* of like nature too (and we referred to the view of liberalism with respect to this in Chapter 4). Ours is the age that is the life of the man referred to in the Introduction to this book who is now old, called modernity, and who looks back over his life and realizes he put work (progress) before family (society and nature). Ours is the age of an almost fully desacralized and exploited world because: a) nature is no longer necessarily even given by the divine, and b) there are now clever people (clever *stupid* people, as indigenous peoples would say) who invent philosophies which say the first (naturalism) and invent technologies that do the second (exploit and degrade).

If in the child age rulers served the people, and in the youth age they served themselves, in the young adult age governments simply serve the pursuit of wealth which, in the mind of the average citizen, is reasoned as both right and natural (this links with the dominant school of economic theory). But as this develops, the rich turn from serving the state to serving themselves, with governments eventually serving them (such as we have today). Ours is the age of the view that we haven't got a place, a family, and a purpose in the universe (this being the materialist worldview). Ours is the age of religious doctrines and humanist ethics (and today, knowledge of ecocide) as the only barriers to behavior which is not completely ignoble. Ours is the age of forgetting what government is about (imitating the One), forgetting what economics is about (household management), and of the Mandevillian fallacy that private vice can produce public virtue.

Ours is the age of the view that what government is about is going with the flow of the selfish, greedy, competitive, and parochial creature which an experimental philosophy (modern science) tells us is a human being. This knowledge means we can sleep well knowing that capitalism is not, or is not completely, ignoble, and we tell ourselves we are civilized and enlightened in our "rational self-interest" (we are more civilized than we were in the youth age, but from the perspective of the mature adult age we are barbaric and unenlightened). Ours is the age of the "myth" of the Great Return, replaced with the "fact" of horizontal evolutionism. The youth age society was still an agrarian society in which the marketplace was not the locus of society. In the young adult age this changes, and the new consciousness allows increasingly for a market fundamentalism and gross industrialization, leading, as many recognize today, to the breaching of societal and environmental boundaries.

The Mature Adult Age

In the mature adult age (considering this in its fullness — which lies some time ahead of us) there is a developed trans-physical self-sense. We are talking about a self-sense *as consciousness itself*: something that is spatiotemporally nonlocalized. Previously, there was not this developed self-sense, hence: a) the view that the universe is objectively real and that everything is just matter/energy, and b) the view that consciousness comes from the brain — leading AI scientists and popular culture to imagine artificial sentient beings and even uploading human consciousness into machines. The mature adult age understanding is different: *there is no physical universe/matter existing independently of consciousness*. The brain comes from consciousness, not the other way around. What the yogi Sri Krishna Prem wrote in 1958 is clear to the average individual in the mature adult age:

> It should be clear from introspective meditation that all forms are sustained in consciousness ... It is the old story of looking for one's spectacles when they are on your nose ... There is not the slightest reason whatever for supposing that anything whatever, physical or mental, exists or can exist save as the content of consciousness.[5]

In the youth age, a developed emotional self-sense meant the self could be selfish and greedy. In the young adult age, added to this we had a developed intellectual self-sense and a rational mind which meant this self-orientation could be rationalized (made out to be natural) and legitimized (both legally and as a life stance). In the young adult age, the world was fully objectified and externalized with the self as an object in that.

We thus had, on top of the notion of a "natural rightness" to capitalism (the "obvious and simple system of natural liberty," as Adam Smith put it), the creed of libertarianism. Libertarianism includes the fundamental right to property ownership, starting with one's body and extending to all private property.

In the mature adult age, with the new reality-sense, there is what we might call a "trans-libertarianism" instead (that's "trans" as in transpersonal). We see this emerging in today's positive transformation movements for social justice, for global democracy, and for a more holistic worldview. For libertarians, individuals are ontologically primary (not some organic whole in which the individual is a part). Personal freedom is therefore qualified as much by freedom *from* society as it is by freedom *in* society — "leave me free to do my thing with my things." Government, for libertarians, is to be small, especially when it comes to taxes and regulations — big only as required to protect this freedom. But for the new trans-libertarians, individuals are not ontologically primary — it *is* an organic whole in which the individual is a part. This is their appreciation, and theirs is therefore a holistic, not a reductionist, individualism. Or to put it another way, theirs is a natural and not merely ideological communitarianism.

The "free society" required by trans-libertarians is therefore different. It excludes freedom *from* society while it still champions freedom *in* society, and it permits (indeed is attracted to) a new collectivism such as in worker-owned businesses and community-owned social enterprises. Government, trans-libertarians understand, can only be small when the private sector has equally stepped up to the plate and assumed its societal responsibility (which it does in the mature adult age). For a government needs less money for spending on the public good when this is also being provided by the private sector. Private virtue produces public virtue. If one wants low taxes, one has to put the flourishing of all selves before the prospering of one's self. This is what mature adults do.

In the mature adult age, there is a return of metaphysics and metahistory. The default view of the universe is not materialist (as it was in the young adult age even if the person was religious) — it simply can't be anymore. Materialism and dualism are appreciated as belonging to the previous age. What is valuable to the individual now, linked with a developing knowledge of interconnectedness, is decreasingly the material and increasingly the nonmaterial and relational. The ideal of monetary wealth for the separate self is replaced by the ideal of prospering together and in ways that count. The economic system of the mature adult age is exchange *for societal and environmental wellbeing*. Governments don't serve the pursuit of wealth anymore but rather this end, and parents don't teach their children they want to make money anymore but that they want to do good in and for the world.

Mature adult age individuals appreciate that the enlightened self-interest in Smithian thought can only take one so far (as can rational thought generally). They have a new enlightened selflessness now (not a more enlightened selfishness): they appreciate that self-interest (and deeper self-identity) lies in the common good — lies in the common *life*. We in the late young adult age would see something like a souped-up Nordic social democracy with three strange features: 1) the whole economy, public *and private*, is directed at achieving the end of societal and environmental wellbeing; 2) the emphasis is not on *ownership* of property (and the legal right to that), but on *use* of property (and the legal *responsibilities* of that); and 3) the money system is in public hands, subjugated only to whatever society needs it for, and is waning.

The "children" (private economic actors) may still, and do still, own parts of the "island," but, through the education system for one thing, and through laws that confirm society's expectations for another, the "parents" (governments) ensure overall that there is successful household management. The children — knowing this is good for them too — go along with this (with necessary "encouragement" as required to behave properly through the laws). We would see, for instance, much stricter, wider, and enforced environment, landlord-tenant, and labor laws, with penalties for breaching these extending to total confiscation and re-operation of the company's assets (taking away the children's "toys"). The complaint of today's nongovernmental organizations (NGOs) is that there are no mechanisms for bringing to heel land-grabbing, nature-polluting, and sweat-shopping corporations; well, in the mature adult age there are, with corporations having a fiduciary duty to society, not to stockholders, and with bad behavior being punished while good behavior is rewarded.

We might reflect on this punitive aspect of things that Article 17 of the Universal Declaration (and the Fourth Amendment to the US Constitution) allows for such confiscation so long as it is not arbitrary/unreasonable. If today's Christian Right remembered the Golden Rule in the Bible — *do to others as you would have them do to you*, and that this was *all the Law* — they would have no problem with this. Pope John XXIII also reminded Christians that "the right to own private property *entails a social obligation*."[6] We might say for balance that the Atheist Left today errs when it throws the baby of God and gnosis out with the bathwater of religion and spirituality, no matter how much blood the Church historically has on its hands. The Golden Rule for humanists is contained in Article 1 of the Universal Declaration: "All human beings are born free and equal in dignity and rights. They are endowed with reason and conscience and *should act towards one another in a spirit of brotherhood*."

The mature adult age is not an atheist age, nor is it a socialist age — although it is an age where a society takes care of its people and its world and is democratic. This is under and courtesy of a new sense of reality and corresponding *nonmaterialist* worldview. Capitalism as a life stance and as an ideology relies on a materialist worldview and ultimately a more juvenile type of consciousness. It therefore naturally dies out in the mature adult age (as does the old Right and Left as we know it). The mature adult age could be seen to be the age of "Buddhist economics," where the society features: a) the pursuit, until achieved, of a universal decent (meaning adequate and modest) standard of living; b) general ethical maturity (not as the exception but as the rule, and not as an obstacle to getting on in life but as the ticket to doing so); c) economic performance conceived as well as measured in terms of wellbeing; and d) living nonviolently — which includes not harming the present generation by allowing inequity and poor living conditions, and not harming future generations through this and further environmental degradation.

> A post-materialistic management paradigm is emerging and characterized by frugality, deep ecology, trust, reciprocity, responsibility for future generations, and authenticity. Within this framework profit and growth are no longer ultimate aims but elements in a wider set of values. In a similar way cost-benefit calculations are no longer the essence of management but are part of a broader concept of wisdom in leadership. Spirit-driven businesses require intrinsic motivation for serving the common good and using holistic evaluation schemes for measuring success.
>
> — Bouckaert and Zsalnai[7]

The mature adult age businessperson sees the bottom line as *maximizing collective wellbeing, not individual profit,* and would relate to the post-materialistic business view as identified by

Luk Bouckaert and Laszlo Zsalnai, directors of the European SPES Institute, who are quoted above. The mature adult age individual has an aversion to greed, opulence, and consumerism, and understands that global economic management does mean, and does require, household management by an "authoritative parenting" type of world government which does exist in the mature adult age. The economic system of the mature adult age is not ideology based — a mere ideology of "wellbeingism" — just as the economic system of the youth age wasn't ideology based. In both cases the new consciousness compelled a new sociopolitical order and that was that. Each age has its own "there is no alternative." The age we are leaving is the one that says that capitalism (or more accurately "wealthism") has no alternative. The one we are entering is the one that says that wellbeingism (whether we call it this or not) has no alternative.

What this means for us today, at the outset of the mature adult age, is that it is about going with the new consciousness, or going with the positive transformation movements, or simply being more mature — all of which amount to the same thing. It means *finally letting go of materialism and capitalism*. Both are being clung on to today, both intellectually and systemically, but we need to let them go to be freer and wiser — to grow up as we are meant to. "Going with" translates as resolving to have, and then making it so, whatever new or improved political structures, policies, and programs are required to achieve global wellbeing. It sounds very simple and in reality it is. We have sustainable development plans today, but we do not yet have *wellbeing* plans. We have development plans but with no real understanding of *what* is to be developed and *why*.

Moreover, these sustainable development plans are handicapped by the young adult age sacred cows and capitalist principles of economic growth, liberalization, and free trade. So a summary of these three would be: a) let's all endlessly earn and spend more, produce and consume more, as if that was the meaning of life, the key to happiness, and ecologically sustainable (none of which is true and, as we have seen, a country's economy may grow without that necessarily meaning the lives of the common folk in it are enriched); b) more things being owned by the "children," who say "leave me free to do my thing with my things" as if that was a good thing for the planetary island (which it's not); and c) more of a society built around the marketplace and market values with minimal government intervention (which is ignoble and makes markets the government — plus we recall Joseph Stiglitz from Chapter 4 saying that free trade deals are managed for the benefit of corporate interests). On the topic of free trade, the former UN Assistant Secretary-General Robert Muller provocatively wrote:

> There is much pressure from the rich countries for free world trade, because they have a distinct historic, economic, scientific and technological advantage to invade the poor countries with their products and advanced marketing and advertisement, often changing the traditional, more healthy, natural, better habits of these countries, where moreover advertisement is very cheap. The poor countries should raise the issue of free migrations, of the freedom of people to settle anywhere on our planet. People should claim this as a fundamental human right. Why only world free movement of goods, and no free movement of people? The United Nations must hold a world conference on the free movement, migration and settlement of all humans on this planet. It will have to be raised sooner or later. The sooner we do, the better.[8]

There is a different moral sensibility in the mature adult age. This includes a universal decent standard of living as something that must be honored — there's no moral ambivalence about this. If I have no existential ties to you, my neighbor — which is how the young adult age sees it — it is a case of "every man for himself" (and every nation for itself). This does not preclude good neighborliness or charity, but it does legitimize *conditional* giving. This is particularly the attitude of the Right today and means that welfare claimants have to look for paid work in return for meagre handouts and a look of scorn, and developing countries have to open up their economies and mostly pay back loans — there's no "something for nothing." The general thought here being that one is freeloading if not engaged in making wealth for somebody. The Right should take another look at Articles 22 and 25(1) of the Universal Declaration.

But if I do have existential ties to you, my neighbor — which is how the mature adult age sees it — it is more a case of "we're all in this together." Already this leans more to the attitude of the Left; however, this is more than just a progressive solidarity view — it is the encroachment of a "we are all one family" esoteric appreciation (as is the spirit behind the socialist movement generally, even if not generally recognized by the Left). This does not preclude conditional giving altogether, but it does preclude conditional giving when it comes to ensuring a universal decent standard of living. This may well translate into providing a universal basic income; however, the important thing is not the specific policy, but the purpose behind *all* policies. A wise government provides an "unstinting supply of the means of life," as Iamblichus reminded us. This giving is essentially the same kind of giving as a parent does to a child — so if it is paternalistic it is also maternalistic, and it is also good sense, for "everything that is honored flourishes."

The mature adult age has a relation to the youth age, as the old adult age has a relation to the child age (and the pensioner age has a relation to the infant age, and the postmortal age has a relation to the prenatal age). If the youth age was the age of organized and separate religions, the mature adult age is the age of personal spirituality and ecumenicalism. The mature adult age is not anti-religious; it is pro-spiritual. It is secularist, but it does not separate the spiritual from the political where: a) we are referring to the spiritual aristocracy principle which, as was said in the previous chapter, is compatible with a democratic system; and b) we are distinguishing the spiritual from the religious. If the youth age was the age of increasing social stratification and warring kingdoms, the mature adult age is the age of class erosion and peaceful international relations.

The mature adult age may be the age in which there are developed technologies (such as robotics and nuclear fusion) that assist the achievement of the wellbeing end, but these would not have been the cause of that end, nor of the system that brought it about (for that was to do with a new type of consciousness coming in, not a technological or socialist revolution). Science and technology will be put to use in the mature adult age (i.e. in the decades and centuries ahead) to restore the damage done to communities and places in the youth and young adult ages. A future historian might characterize the mature adult age as the "Great Restoration," and link it with the community-service *Vanaprastha* stage of life in Hinduist philosophy.

The Old Adult Age

The whole business of man is the arts.

— William Blake

The whole business of humankind in the old adult age is the arts, after the Great Restoration. We are talking about a planetary society that has, through the mature adult age, and as a whole, *organically* de-grown, de-industrialized, and re-socialized (i.e. moved from having public and private sectors into having just one "social sector" — which is to say moved forward in the cycle back to how it was in the child age, but with the learning that has come from the journey). And we are talking about: a) arts that express the Good, and b) "arts" having the broadest meaning, that is, referring not just to the arts and crafts but extending to architecture and urban design and indeed all occupations.

In the mature adult age, there was the beginning of a trans-*emotional* self-sense: here in the old adult age it is developed. The original Arcadia was serene because there was not yet a developed emotional nature, and thus not yet sufficient non-group orientation. In order for there to be a new Arcadia again, there needs to be a sufficient non-self-orientation again. There needs to be a transformation not just in the *mind* in terms of a new understanding of the universe but also in the *heart* too. The center of *personality*, of selfism, being the emotional nature. We are talking about the requirement for a deeper spirituality / transpersonal development, and a deeper unicity as a result, manifesting for one thing in a demilitarized and borderless world.

"What is mine is also yours should you need it" featured in the child age and will do so again in the old adult age. In the former, this was based on a sense of property strong enough to take responsibility for that property – without yet a concept of ownership. In the latter, this is based on a degree of unpossessiveness strong enough to rise above the concept of ownership, while still retaining responsibility for the property. In the former, there was a symphony between the land being the people's, and the people being the land's. In the latter there is the same symphony; however, it is not based on pre-intellectual nature religion, but rather on a suprarational knowledge of interkingdom connectedness, taking in the Great Cycle as referred to in Part I.

In the child age, rulers held the sacred trust to be guardians of the land, the people, and the overall "religious" culture. In the old adult age, a planetary government will do an equivalent, which translates as delivering on all three priorities of government. To recall, these three were: 1) seeing to it that the basic needs of all citizens are met; 2) seeing to it that there is good universal education, the primary aim of which is the production of people who possess rectitude, an altruistic nature, and cosmopolitan identity, are geopolitically and historically informed, and are intellectually developed; and 3) seeing to it that there are cultural activities, including everyday work, which utilize people's education and natural talents in service to the wider good. This service being "artistic" in its performance. It is interesting to note how in a lot of science fiction we can already imagine such a world.

The child age society was a non-growth, subsistence-based society, and was moneyless. The old adult age society will equally be non-growth outwardly (but not inwardly) and moneyless. And the old adult age society will be equally oriented toward basic needs, but that doesn't mean a return to an agrarian and non-technological society — reunited with nature in the heart and mind, yes, but we can imagine it being far more technological than our own or even the mature adult age (although with technologies dedicated to societal and higher purpose). In the old adult age, capitalism and nationalism, together with money and borders, are past concepts and past realities, hard to understand; just as in the child age they were future concepts beyond understanding. The "local community" in the old adult age applies to the local community anywhere and the global community everywhere.

The old adult age society is a contemplative society. Contemplating on the cosmic order, that would be, which would also be contemplating on the self and its experiences and expressing that contemplation in cultural productions. Productions that don't have this contemplative signature, or in other words spiritual meaning, are not useful productions where the orientation of people is more to the other-worldly than to the this-worldly (as we in the young adult age would see it — for those in the old adult age there is no distinction between the two, and there is no word for "spiritual"). We are recalling the *whole* of nature now (the horizontal and vertical universes) in Part I, and Huston Smith's reference to a "right angle turn" to the qualitative. In a sense, the mature adult age was about sorting out the quantitative, with just the beginning of a focus on the qualitative; now in the old adult age we focus more on the qualitative, and we fully focus on the qualitative in the succeeding "end of politic action" pensioner age.

One of the biggest concerns today is what the United Nations calls the lack of decent and non-precarious jobs for all — a situation getting worse under neoliberal economic globalization and especially worrying as we enter the "fourth industrial revolution" of AI and cyber-physical systems. Most would agree that a decent job is one where the person can be creative in what they do or make, and where the work is not just rewarding but also vocational. It is the saddest thing not to be doing something worthwhile and meaningful in one's working life. To put it in the urban slang, as it is fitting to do, there were no "shit jobs" in the child age, and there will be none again in the old adult age. For by then, we will have well and truly put behind us the current economic system and reversed the trajectory begun in the youth age from artisanship to mass production, and from work which was *cultured* (noble, elegant, beautiful) to work which was/is, for a great many today, just work and uncultured.

Chapter 6

The Way Forward

Introduction

A successfully run planet would be one where the civilization's "management model" imitates that of the universe (as above, so below). The universe's management model is to provide for autonomous existence but with a built-in return magnet, to provide for growth to selfhood but then to union with one's author, and to further express the Good with each "business cycle." Translated, we might say it is to provide for free will and liberty, but for that free will and liberty to be knowingly, happily, and creatively deployed – time and time again – in doing good for and as part of the wider whole. What we need today more than anything else is to move from an overall policy goal of economic growth to an overall policy goal of societal and environmental wellbeing. The mental shift required is great, *but it is the solution to the current economic and environmental crisis.*

It is also the solution to the current political crisis, where we are referring to the liberal rules-based international order under threat from illiberalism, authoritarianism, and nationalism. The problem is not with the "liberal" part of that order per se, nor with the "international" part per se, but with: a) the liberal part including allowing the private sector – the "children" – to run riot over the planetary island; and b) *mere* internationalism, not also global government of the whole island. Freedom to be selfish and greedy – to put private profit before people and planet (which always produces illbeing somewhere) – is no longer acceptable; freedom to be non-selfish and non-greedy (which does not preclude modest profit-making) is acceptable. And countries merely working together is insufficient. We have local economies and local governments. We have national economies and national governments. We have a *global* economy but not yet – and this is required – a *global* government.

This would be going forward, not backward. The future is not a return to the "golden age of capitalism," much as many reactionaries pine for this. The future is not each country climbing into its own bunker like a survivalist. The neo-nationalism we have is as a result of the *typ*e of globalization we have (neoliberal). Change the type of globalization and things are very different. The cry for national renewal based on higher living standards for ordinary people is understandable and legitimate. But this does not require the un-pooling of sovereignty and a retreat from supranationalism. On the contrary, we must recognize that national legislatures which prioritize economic growth inevitably end up serving the global "nobles," and therefore not serving the wellbeing of ordinary people – not giving them what they want and need. The answer to our global problems is constraining the liberal (so businesses have to operate in a socially and environmentally responsible way) and releasing the international (so proper and actual global governance can arise through the United Nations).

The future is not the dictatorship of the proletariat. The future is not the Age of Aquarius. But it can be looked at this way: If we were to take the dialectical materialism out of socialist thought, and the romantic mysticism out of New Age thought (thus leaving merely the core concern with a universal decent standard of living and the delivery of that on the one hand, and the core recognition of an extra-scientific reality and the implication of that on the other hand), then this is the future. We may entertain ourselves with cyberpunk dystopias, but they are just that — fictions. The Magna Carta, the Slavery Abolition Act, the Universal Declaration of Human Rights ... the trajectory is toward a more enlightened and humanist (as in concerned with the welfare and dignity of all humans) society, achieved through mentally maturing *beyond* capitalism, materialism, and nationalism. If one were to draw a depiction of the future in the form of a person, it would be a 50-year-old multi-ethnic Swede who practices raja yoga, reads Joseph Campbell, and works as a UN special rapporteur.

The Global Democracy Manifesto calls for "deeper forms of democracy." This book supports that call and says that the future (which is to say the incoming generations) demands that we add the compatible principles of meritocracy and spiritual aristocracy (as defined) to a global democratic system. Global democracy is the key to UN reform. The manifesto calls for a "new paradigm of development"; this book says that paradigm, in a word, is wellbeing (but says: see also the "mature adult age" in Chapter 5). The beauty of "wellbeingism" is that it does not rely on socialization/nationalization and it is easy for the public to understand: it is not socialism, nor is it capitalism, nor is it even centrism (although it could be seen to be a "higher evolution" of social democracy). It is based on the pursuit of wellbeing rather than wealth. It is not statism as opposed to the free market; it is the state supporting the private sector (through laws) to do the right thing, complemented by a wiser view of the universe and an improved and global democratic order.

The manifesto calls for a "globalisation of the political institutions responsible for their regulation and control." "Their" refers to the currently unmanaged or poorly managed global processes and systems — including particularly the international monetary and financial system — and transnational corporations. This book supports that call, offers a picture of a family shipwrecked on a desert island for political (and corporate) leaders to reflect upon, and says that the future demands world economic management as world household management. This book says that world economic management is best done by, and in the end can only be done by, a world government – not merely by national governments acting more closely together, and certainly *not* by some multi-stakeholder governance model which includes corporations (an idea entertained at the World Economic Forum). We already have a multi-stakeholder model of government – it's called democracy.

This book says that a future world government should see its role as "shepherd to the human herd," as Plato said. This would be a guardian and directing role which nurtures, protects, and honors human members of society, as Iamblichus said. This "Platonic" world government would not be one that is remote from the people – it would be the people's own

Table 2: Wealthism and Wellbeingism

"Wealthism" (links with a materialist worldview)			*"Wellbeingism" (links with a nonmaterialist worldview)*
Capitalism	Socialism	Centrism	
Private ownership of wealth and economic growth	Public ownership of wealth and economic growth	Private and public ownership of wealth and economic growth (always tends to plutocratic capitalist dominance)	Private and public ownership of wealth, but with legal obligation to use capital responsibly, i.e. to enhance societal and environmental wellbeing

government, with a recognition of the to-be rule of the wise and a reaching toward that by voters and the elected alike, and the to-be election of rulers (and removal if they're no good) by the people.

The manifesto calls for a "wider and much better coordinated structure, which involves democratic regional institutions on all the continents, the reform of the International Court of Justice, a fairer and more balanced International Criminal Court and a United Nations Parliamentary Assembly as the embryo of a future World Parliament." This book supports that call and: a) describes a subsidiarity-based planetary government structure with poleis at global, regional, national, provincial, and local levels; b) recognizes the unitive regional ambitions of the African Union and the Union of South American Nations following the EU model – and supports continentalism generally; c) recognizes that the World Bank, International Monetary Fund (IMF), and World Trade Organization (WTO) must be under the UN (but that goes back to UN reform generally, and a comprehensive proposal for this has been recently advanced[1]); d) supports the idea of a standing reserve of UN military personnel, which should be a no-brainer (it is estimated that a force of 300,000 would cost around $25 billion a year, which is just 1.25% of current global military spend); and e) envisages a world parliament evolving in due course into a full "Platonic" world government. Former UN Secretary-General Boutros Boutros-Ghali wrote on the UNPA proposal:

> The establishment of a Parliamentary Assembly at the United Nations has become an indispensable step to achieve democratic control of globalization. Complementary to international democracy among states, which no less has to be developed, it would foster global democracy beyond states, giving the citizens a genuine

> voice in world affairs. As the Campaign's appeal rightly implies, a United Nations Parliamentary Assembly could also become a catalyst for a comprehensive reform of the international system. In particular, I would like to point out, it should become a force to provide democratic oversight over the World Bank, the IMF and the WTO. We cannot just dream or wait for someone else to bring our dream about. We must act now.[2]

The Global Democracy Manifesto implicitly recognizes, and this book explicitly recognizes, the following words by E. F. Schumacher (and would merely swap the word "moral" for "psychological"):

> The generosity of the Earth allows us to feed all mankind; we know enough about ecology to keep the Earth a healthy place; there is enough room on the Earth, and there are enough materials, so that everybody can have adequate shelter; we are quite competent enough to produce sufficient supplies of necessities so that no one need live in misery ... There is no economic problem and, in a sense, there never has been. But there is a moral problem.

This book would suggest that the notional extraterrestrial would patiently hear us describe all the political obstacles in the way, including the private financing of electoral campaigns in the US and corruption in many countries, but say: "Yes, I hear you, but all these obstacles are man-made so can be changed — so, I ask again, what is your problem?" Our problem is not so much "out there" in terms of structures and systems as "in here" in terms of our consciousness and worldview.

Wellbeing, Not Economic Growth

Consider the wealthy businessman who goes on a course at an ecovillage, has an epiphany, and subsequently reorders his life. He pursues wellbeing now, not year-on-year earning and spending more. He is a reformed capitalist now — he follows the Golden Rule in his business practices, which includes treating his staff well (doing to them as he would have them do to him). He is mindful of his actions and leaves a small carbon footprint. He is generous to those less fortunate than himself and is more open and forgiving with colleagues. He withdraws from vices such as gambling, and he commits to wholesome earning and spending. He eventually moves to the country to be closer to nature, and goes part-time to spend more time with his family and to do things that are unpaid but make him happy — things that he did at the ecovillage such as gardening, crafts, and community work. His income goes down and it stays down, but he doesn't care anymore: he has greater wellbeing. He says to himself: "As long as I have enough for my needs, that is enough."

Looked at as a global whole, we not only have more than enough natural resources to go around but also more than enough wealth resources. Looked at as a global whole, there would be no problem with going "part-time." We may frame what the wealthy businessman does as essentially the planetary civilization trajectory to the old adult age. There is no reason why the wealthiest nations today, *together* (it would need to be together — and the EU could lead on this), cannot begin this reordering trajectory now. Following the Golden Rule would, for one thing, translate as taking steps toward new world domestic laws (not least, those relating to labor standards). Being generous to those less fortunate than oneself has in mind higher and free official development assistance, particularly for the least-developed countries and particularly for vital infrastructure (just living up to official development aid [ODA] targets would be a start). And being more open and forgiving with colleagues has in mind intellectual property and debt-forgiveness.

Withdrawing from vices such as gambling specifically has in mind the global casino (the stock market), which is not a necessary part of a monetary economy — while we still have a monetary economy. Committing to wholesome earning translates as having progressive tax policies, and new taxes as required, on environmental spoliation and unearned income/the rentier economy (these being only right and fair under a trans-libertarian or more mature way of looking at things). Wholesome spending particularly translates as higher spending on public services and the public estate. The widespread unemployment and underemployment problem of today could be solved by creating a new global program of jobs aimed at the "Great Restoration" of communities and places (such work including the protection of the global commons and the enforcing of the new world domestic laws). The young of today want to work, but they want to be engaged in something meaningful and satisfying, not just shrivel their souls in corporate servitude.

A universal basic income could be part of the mix, as said, and a national and global accounting based on wellbeing would be part of the mix. The wealthy businessman doesn't require himself to grow economically, so we would need to ensure that we have a money creation and lending system that doesn't require economic growth in order to service debt. A public takeover of banks isn't required (but neither would it be unacceptable). A public takeover of *the system* so that it works for society and not against it is required. The economist David Korten wrote on this subject:

> A nation's money supply is created by either government's spending new money into existence or a bank's loaning it into existence. The former approach allows a government to pay for public services beyond the amount of its tax revenues. The latter generates large profits for private bankers. Though it's not generally acknowledged by the public, virtually all money is now loaned into existence by private banks, which means a nation's money supply and the stability of its money system depend on continuously expanding debt to create enough money to repay the old debts and avoid bankruptcy — a primary reason why capitalist economies are prone to collapse if they do not grow exponentially. Placing a 100 percent reserve requirement on demand deposits in the banking system and returning the function of creating national currencies to government would largely eliminate the federal government's need to borrow, reduce the power of the banking system, and eliminate an important source of the money world's growth imperative.[3]

None of these solutions/recommendations appears inconsistent with a summary of the situation, and the need, taken from a recent UN report on social justice:

> Income-related inequalities, notably in the ownership of capital and other assets, in access to a variety of services and benefits, and in the personal security that money can buy, are growing. There is also greater inequality in the distribution of opportunities for remunerated employment, with worsening unemployment and underemployment in various parts of the world affecting a disproportionate number of people at the lower end of the socio-economic scale. The inequality gap between

the richest and poorest countries, measured in terms of national per capita income, is growing as well. The popular contention that the rich get richer and the poor get poorer appears to be largely based on fact, particularly within the present global context ... The concept of reform, so often invoked in recent years to facilitate economic deregulation and privatization, could be constructively applied by liberal democracies and other regimes inspired by liberal principles to identify the requirements of social justice and implement appropriate policies. To an extent, the United Nations, with its efforts to strengthen the role and contribution of civil society, is taking the lead and paving the way for international and global democracy, a prerequisite for global social justice. Social justice is not possible without strong and coherent redistributive policies conceived and implemented by public agencies. A fair, efficient and progressive taxation system, alluded to in Commitment 9 of the Copenhagen Declaration on Social Development, allows a State to perform its duties, including providing national security, financing infrastructure and public services such as education, health care and social security, and offering protection and support to those who are temporarily or permanently in need. While the scale of taxation and public obligations varies widely according to the wealth and capacity of each country, it is generally accepted that greater financial participation is required from those with more resources at their disposal. This constitutes a sacrifice that in well-functioning liberal and social democratic societies is accepted as part of the social contract binding citizens together. Official development assistance (ODA) to poor and developing countries is a manifestation of redistributive justice at the international level, and various proposals for taxes on global transactions derive from the same objective of

> promoting fairness and solidarity. Redistributive ideas and practices are currently under attack. Governments and international organizations vacillate between the adjustment, neglect and abandonment of redistributive policies, but there is no evidence yet of any socially, politically or economically viable alternative strategies.[4]

Beyond the 2030 Agenda

This book acknowledges that there has been progress in political thinking and global action in recent years, and wholeheartedly supports the 2030 Agenda and the Wellbeing Economy Governments initiative.[5] While we move to implement the 2030 agenda, though, we can go further in thought and future planning. We have gone from "it's all about economic growth" to "it's all about inclusive and sustainable economic growth." The next stage is "it's all about wellbeing." Here's an idea: a longer-term (say to 2070) World Wellbeing Plan — one that is legally binding — succeeding the non-legally binding 2030 Agenda. If we need to be reminded, economic growth does not equal wellbeing, nor does economic growth necessarily lead to wellbeing. A "strong" economy can be a decidedly *unhealthy* (leading to illbeing) economy. The Canadian Index of Wellbeing (CIW) puts it very well:

> There is a *feeling* that all is not well in Canada. But it's more than a feeling; it's a fact. When we compare trends in the wellbeing of Canadians to economic growth in the period from 1994 to 2014, the gap between GDP and our wellbeing is massive and it's growing. When Canadians go to bed at night, they are not worried about GDP. They are worried about stringing together enough hours of part-time jobs, rising tuition fees, and affordable housing. They are thinking about the last time they got together with friends or the next time they can take a

> vacation. Maybe that's why we are getting less sleep than 21 years ago ... GDP rests on the philosophic assumption that all growth is good — a rising tide lifts all boats [it is falsely claimed]. But is all growth really good? And are all activities where no money changes hands of no value? If you're talking about GDP, the answer to both questions is "yes". GDP makes no distinction between economic activities that are good for our wellbeing and those that are harmful. Spending on tobacco, natural and human-made disasters, crime and accidents, all make GDP go up. Conversely, the value of unpaid housework, child care, volunteer work, and leisure time are not included in GDP because they take place outside of the formal marketplace. Nor are subtractions made for activities that heat up our planet, pollute our air and waterways, or destroy farmlands, wetlands, and old-growth forests. The notion of sustainability — ensuring that precious resources are preserved for future generations — does not enter the equation.[6]

The CIW defines wellbeing as

> the presence of the highest possible quality of life in its full breadth of expression focused on but not necessarily exclusive to: good living standards, robust health, a sustainable environment, vital communities, an educated populace, balanced time use, high levels of democratic participation, and access to and participation in leisure and culture.

There are other indexes with similar definitions. We could add this to — or merge this with — the UN definition of human development which is about allowing people to live a "long, healthy, creative life, and to enjoy a decent standard of living, freedom, dignity, self-respect and the respect of others." We make especial note that wellbeing is not reducible to having disposable income to spend at the mall: it incorporates aspects of nonconsumerism and self-actualization in the Maslowian sense. It is the neoliberalists today and not the anti-capitalists who need to be put on the witness stand and made to answer two key questions:

a) So what's wrong with running your businesses in a socially and environmentally responsible way, that is, putting people and planet before profit?
b) If you say we need to give you more freedom to give us greater growth, but that growth is not giving us wellbeing, why should we give you more freedom and why should we focus on growth?

Five Common but Mistaken Economic Views

1. "It's all about economic growth and individual freedom, with governments pursuing the first and honoring the second." No, it's all about societal and environmental wellbeing, and individual freedom to serve (but not disserve) that, with government pursuing the first and ensuring the second.
2. "Economic growth is the way to reduce poverty." No, household management is the way to reduce poverty. Poorer countries should be seen as areas of deprivation within the one global "country" that is the planet. We are all the same family.
3. "Economic growth is best achieved by supporting the investors and the entrepreneurs who are the wealth creators." No, even if it was about economic growth — which it's not — the wealth creators are the workers and the carers, and it is they who need supporting. But to be

really clear: it's not about economic growth; it's about wellbeing. Think of that post-epiphany businessman. *Think of one's own life.*

4. "More free trade is what we need." No, more trade conducted by private economic actors for the purpose of enhancing societal and environmental wellbeing is what we need. Think of a monastic community. Does it matter whether the land used to grow the community's food is legally in the name of some or all of the monks — does this affect whether everyone in that community gets to enjoy the food? No. It's not about private or public ownership; it's about the requirement, accepted and understood, to use the capital to benefit the whole community. Intracommunity, are the monks all busy earning and spending more year after year — is that what their life is about? No. Their lives are about something higher and nobler, ultimately about self-improvement in a cosmic context. Now transplant this model-cum-mindset into our society — as the path forward out of the current mess. What is required are global laws requiring businesses to act in a socially and environmentally responsible way. What is required is a new cultural and political focus on human development and wellbeing. As said, the next stage is: "it's all about wellbeing."
5. "Economic growth is required to fund public services." No, taxes are required to fund public services (or ordinarily correlate with government spending on them) where there is a non-sharing society (there are no taxes in a hunter-gatherer society). Consider two Amazonian tribes. In one, the tribe shares the food the members gather. In the other, the members insist on owning the food they gather for themselves. The chief (government) of the second tribe then has to "tax" the members to ensure everyone in the tribe eats. The lesson is this:

selfish greed requires high taxes; selfless, public-serving behavior enables low taxes. *Because* people think it's all about economic growth, the debate is limited to whether lower taxes stimulate growth, or higher taxes impede growth, and to a lesser extent on whether economic growth necessarily leads to more tax revenues to spend on public services (which it doesn't). And because the people currently winning this debate are those who say that lower taxes do stimulate growth, and that economic growth does lead to more tax revenues, we have the common mistaken view that supporting capitalism is good for all or at least a necessary evil. It is neither. The wealthy businessman's epiphany — a breakthrough in lucid thinking — is that it is all about wellbeing, not economic growth. The new debate a post-epiphany society has is what is required to achieve societal and environmental wellbeing. One thing that is required is a global household management approach. Another is the right tax policies, as alluded to. The trans-libertarian understands that a government needs less money for spending on the common good when this is also being provided by the private sector. *The private sector is to publicly serve, not the other way around.* Think of that shipwrecked family and the children who are to grow up (under good parenting) to add to the wellbeing of the family and the island. Think of how they are not focused on economic growth, but on flourishing into the best people they can be.

How do we reduce poverty (created by the wealth-seeking economic system, including in its manifestation as colonialism/neocolonialism, in the first place)? The usual answer is to grow economies — by letting capitalism do its thing. The question is then asked: "But what do we do about capitalism bringing inequity and environmental decline?" The groupthink of today's political leaders is: "Well, we are encouraging corporations to follow voluntary codes and sign up to the UN's Global Compact, and we are trying to tackle tax evasion and support the development of green technologies, but beyond that we need to grow our economies, so we must let capitalism do its thing." But capitalism cannot *do* inclusivity and sustainability – it's just not in its DNA. And we need to move beyond merely *promoting* good behavior to *requiring* good behavior. This goes back to the question of liberty, of course (the freedom to be the ignoble Hobbesian creature or the noble Platonic soul), and to our civilization worldview and the lens through which we see reality.

It is welcome that most governments have committed to Agenda 2030 and the Paris Agreement on Climate Change. It is welcome that some corporations have signed up to the Global Compact. A social business model is a good model: the future of business generally is to contribute to the wellbeing of the human family and the planetary island – not so long as this doesn't interfere with profit-making, but with the bottom line changing from maximizing individual profit to maximizing collective wellbeing (and correspondingly, the legal right to make a profit changing to a legal responsibility to put people and planet first). The required grand shift in thinking includes understanding that we reduce poverty by doing as we would in our own homes – by household management. It's not about growing economies; it's about achieving wellbeing. We achieve wellbeing in the first instance by having it as the global policy goal. Then we take the necessary steps (which automatically means a non-laissez-faire economy) to achieve it.

"Don't hand the keys back to the people who crashed the car" was a line used by the Conservative Party in the UK. They were referring to the financial collapse of 2008 when at the time there was a Labour government in power. Using specious arguments, they managed to convince many voters that the Labour government was responsible for the crash. It wasn't, of course. As the former Financial Standards Authority chairman, Adair Turner, put it: "The financial crisis of 2007/8 occurred because we failed to constrain the private financial system's creation of private credit and money."[7] Borrowing and adapting the line, this book says to today's political leaders: "On behalf of the world's people, and in order to remove the primary cause of the populist authoritarian nationalism that is blighting the international order, take the keys back from the children who are crashing the car."

Global economic management does mean, and does require, global household management. This book provides a vision of things and guiding principles, but courage, leadership, and *unity* is required to defeat the common "threefold enemy" (and it *is* that), namely: a) market fundamentalism, b) an out-of-control – literally – money system, and c) the rapacious neofeudalist "nobles" who have too much power and play one government off against another ("we'll move our investment elsewhere unless you are kind to us"). The message today to political leaders is simple: Listen to the world's people, respond to the need, and follow your own higher wisdom – this is how you run a planet. If this can be the theme of a future United Nations conference and/or a G20 summit, then all the better.

Notes

Introduction

1. "Transforming Our World: The 2030 Agenda for Sustainable Development." https://sdgs.un.org/2030agenda.
2. John Holman, *The Return of the Perennial Philosophy* (London: Watkins, 2008).
3. Glenn Alexander Magee, ed., *The Cambridge Handbook of Western Mysticism and Esotericism* (New York: Cambridge University Press, 2016), xxviii–xxix.
4. Willis Harman, *Global Mind Change: The Promise of the 21st Century* (San Francisco: Berrett-Koehler, 1998), 127–8.
5. Augusto Lopez-Claros, Arthur L. Dahl, and Maja Groff, *Global Governance and the Emergence of Global Institutions for the 21st Century* (Cambridge: Cambridge University Press, 2020), 484.
6. Paul Kennedy, Dirk Messner, and Franz Nuscheler, eds, *Global Trends and Global Governance* (London: Pluto Press, 2002), 8.
7. Pope Francis, *Laudato Si': On Care for Our Common Home* (Our Sunday Visitor, 2015).

Chapter 1: Our Place in the Universe

1. Seyyed Hossein Nasr, "Spirituality and Science." In Seyyed Hossein Nasr and Katherine O'Brien, eds, *The Essential Sophia* (Bloomington, IN: World Wisdom, 2006), 212.
2. Frithjof Schuon, "Sophia Perennis and the Theory of Evolution and Progress." http://www.frithjof-schuon.com/evolution-engl.htm.
3. Huston Smith, *Beyond the Post-Modern Mind* (Wheaton, IL: Quest Books, 1996), 200.
4. Schuon, "Sophia Perennis and the Theory of Evolution and Progress."

5. Alice Bailey, *A Treatise on White Magic* (London and New York: Lucis Trust, 1991), 27–28.
6. Frithjof Schuon, *Light on the Ancient Worlds* (Bloomington, IN: World Wisdom, 1984), 111.
7. Ibid., emphasis mine.
8. Titus Burckhardt, *Alchemy: Science of the Cosmos, Science of the Soul* (Shaftesbury: Element Books, 1986), 51.
9. Iamblichus, *On the Mysteries of the Egyptians, Chaldeans and Assyrians,* trans. Thomas Taylor (London: Bertram Dobell, 1895), 23.
10. Thomas Taylor, *A Dissertation on the Philosophy of Aristotle* (London: 1812), 499.
11. Plotinus, *The Six Enneads,* trans. Stephen Mackenna and B. S. Page, fifth ennead, second tractate, chapter 1. http://classics.mit.edu/Plotinus/enneads.html.
12. Proclus, *On the Theology of Plato,* trans. Thomas Taylor, book II, chapter XI. https://archive.org/details/ProclusOnTheTheologyOfPlato-ElectronicEdition.
13. Manly Hall, *Journey in Truth* (Los Angeles: Philosophical Research Society, 1945), 27–32.

Chapter 2: The Nature of Things

1. Helena Blavatsky, *The Secret Doctrine: The Synthesis of Science, Religion, and Philosophy* (New York: Theosophical Publishing Company, 1988), 329.
2. Quoted in Alice Bailey, *A Treatise on Cosmic Fire* (London and New York: Lucis Trust, 1989), 278.
3. Plato, *Timaeus,* trans. Benjamin Jowett. http://www.gutenberg.org/files/1572/1572-h/1572-h.htm#link2H_4_0010.
4. *Liber XXIV Philosophorum.* The Matheson Trust for the Study of Comparative Religion. http://themathesontrust.org/papers/metaphysics/XXIV-A4.pdf.

5. Iamblichus, *The Life of Pythagoras*, trans. Thomas Taylor, chapter XXIX. http://www.universaltheosophy.com/legacy/works/life-of-pythagoras/.
6. Helena Blavatsky, *The Key to Theosophy* (New York: Theosophical Publishing Company, 1889), 83.
7. Blavatsky, *The Secret Doctrine*, 579.
8. René Guenon, "Sacred and Profane Science." In Mehrdad M. Zarandi, ed., *Science and the Myth of Progress* (Bloomington, IN: World Wisdom, 2003), 33–42.
9. Thomas Taylor, *The Philosophical and Mathematical Commentaries of Proclus* (London: 1792), preface. https://archive.org/details/philosophicalan01marigoog.
10. Alice Bailey, *A Treatise on Cosmic Fire* (London and New York: Lucis Trust, 1989), 1059.
11. Huston Smith, *Beyond the Post-Modern Mind* (Wheaton, IL: Quest Books, 1996), 221.
12. *The Gospel of Truth*. The Gnostic Society Library. http://www.gnosis.org/naghamm/got.html.

Chapter 3: Past-Present-Future

1. Aldous Huxley, *The Perennial Philosophy* (New York: Harper & Row, 1990), 34.
2. Joseph Epes Brown, "American Indian Experience." In Ranjit Fernando, ed., *The Unanimous Tradition* (Colombo: The Sri Lanka Institute of Traditional Studies, 1991), 32.
3. Ibid.
4. Mircea Eliade, *The Sacred and the Profane*, trans. Willard Trask (New York: Harcourt, Brace & World, 1963), 22.
5. Huston Smith, *Beyond the Post-Modern Mind* (Wheaton, IL: Quest Books, 1996), 117.
6. Richard Tarnas, "The Greater Copernican Revolution and the Crisis of the Modern World." In David Lorimer and Oliver Robinson, eds, *A New Renaissance: Transforming*

Science, Spirit and Society (Edinburgh: Floris Books, 2010), 52–3.

7. Brown, "American Indian Experience."
8. Rudolf Steiner, *Occult Science: An Outline*, trans. George and Mary Adams (London: Rudolf Steiner Press, 1969), 108–9.
9. Ovid, *Metamorphoses*, trans. Samuel Garth, John Dryden et al., book the first. http://classics.mit.edu/Ovid/metam.html.
10. Frithjof Schuon, *The Eye of the Heart: Metaphysics, Cosmology, Spiritual Life* (Bloomington, IN: World Wisdom, 1997), 69.
11. Alice Bailey, *The Rays and the Initiations* (London and New York: Lucis Trust, 1993), 660.
12. Plotinus, *The Six Enneads*, trans. Stephen Mackenna and B. S. Page, sixth ennead, ninth tractate, chapter 5. http://classics.mit.edu/Plotinus/enneads.html.
13. Huston Smith, *Beyond the Post-Modern Mind* (Wheaton, IL: Quest Books, 1996), 66.
14. Seyyed Hossein Nasr, "In the Beginning Was Consciousness." In Seyyed Hossein Nasr and Katherine O'Brien, eds, *The Essential Sophia* (Bloomington, IN: World Wisdom, 2006), 205–6.

Chapter 4: Good Global Government

1. E. F. Schumacher, *A Guide for the Perplexed* (London: Jonathan Cape, 1977), 154.
2. Rudolf Steiner, "Inner Aspect of the Social Question." http://wn.rsarchive.org/Lectures/GA193/English/RSP1974/19190309p01.html.
3. Iamblichus, *The Letters*, trans. John M. Dillon and Wolfgang Polleichtner (Atlanta: Society of Biblical Literature, 2009), 17.
4. Sthenidas the Locrian, *On A Kingdom*, trans. Kenneth Sylvan Guthrie. https://www.holybooks.com/wp-content/uploads/The-Complete-Pythagoras.pdf.
5. Iamblichus, *The Letters*, 53.

6. Noam Chomsky, *Who Rules the World?* (London: Hamish Hamilton, 2016), 2.
7. "Global Wealth Report 2022." Credit Suisse Research Institute. https://www.credit-suisse.com/about-us/en/reports-research/global-wealth-report.html.
8. https://taxjustice.net/reports/the-state-of-tax-justice-2023/.
9. Mark Carney, "The Spectre of Monetarism." http://www.bankofengland.co.uk/speech/2016/the-spectre-of-monetarism.
10. Joseph Stiglitz, "The Coming Great Transformation." https://business.columbia.edu/sites/default/files-efs/imce-uploads/Joseph_Stiglitz/The%20Coming%20Great%20Transformation%20Final.pdf.
11. Patrick Bond, "Multinational Corporations Invade Global Governance Institutions." https://www.civicus.org/documents/reports-and-publications/SOCS/2017/essays/multinational-corporations-invade-global-governance-institutions-causing-for-profit-paralyses.pdf.
12. "20 Solution Proposals for the G20." https://www.global-solutions-initiative.org/wp-content/uploads/2020/11/20_Solutions_for-the_G20_17-7.pdf.
13. Global Democracy Manifesto. https://www.opendemocracy.net/en/global-democracy-manifesto/.
14. Plato, *The Republic*, trans. Benjamin Jowett, book VII. http://classics.mit.edu/Plato/republic.8.vii.html.
15. Aldous Huxley, *The Perennial Philosophy* (New York: Harper & Row, 1990), 93.
16. Manly Hall, *Lectures on Ancient Philosophy* (Los Angeles: Philosophical Research Society, 1984), 192.
17. Plutarch, *De Exilio*. http://www.perseus.tufts.edu/hopper/text?doc=Perseus%3Atext%3A2008.01.0308%3Asection%3D5.
18. Preliminary Draft of a World Constitution. https://en.wikisource.org/wiki/Preliminary_Draft_of_a_World_Constitution_1948.

19. Constitution for the Federation of Earth. https://en.wikisource.org/wiki/Constitution_for_the_Federation_of_Earth.
20. United Nations Parliamentary Assembly proposal. http://en.unpacampaign.org/.
21. https:// www.democracywithoutborders.org/29850/international-poll-public-supports-a-world-parliament-and-world-law/.
22. BBC World Service Poll, 2016. https://www.bbc.co.uk/mediacentre/latestnews/2016/world-service-globescan-poll.
23. Plato, *The Politicus,* trans. Thomas Taylor (London: 1804), 174.

Chapter 5: Economics, History, Society

1. G. W. F. Hegel, *Lectures on the Philosophy of History,* trans. J. Sibree (London: G. Ballard and Sons, 1914), 82.
2. Ovid, *Metamorphoses,* trans. Samuel Garth, John Dryden et al., book the first. http://classics.mit.edu/Ovid/metam.html.
3. Marshall Sahlins, *Stone Age Economics* (Chicago: Aldine-Atherton, 1972), 169.
4. John Gowdy, "Hunter-Gathers and the Mythology of the Market." http://libcom.org/history/hunter-gatherers-mythology-market-john-gowdy.
5. Sri Krishna Prem, *The Yoga of the Bhagavat Gita* (London: John Watkins, 1958), 196.
6. Pope John XXIII, *Pacem in Terris,* 1963, emphasis mine. http://w2.vatican.va/content/john-xxiii/en/encyclicals/documents/hf_j-xxiii_enc_11041963_pacem.html.
7. Luk Bouckaert and Laszlo Zsolnai, "Spirituality in Economics and Business." https://laszlo-zsolnai.com/spirituality-in-economics-and-business/.

8. Robert Muller, "2000 Ideas & Dreams for a Better World," Capitalism, Idea 132. http://robertmuller.org/topics/.

Chapter 6: The Way Forward

1. Joseph E. Schwartzberg, *Transforming the United Nations System: Designs for a Workable World* (Tokyo: United Nations University Press, 2013).
2. Boutros Boutros-Ghali. In Andreas Bummel, "Developing International Democracy." https://www.unpacampaign.org/files/2010strategy_en.pdf.
3. David Korten, *The Post-Corporate World: Life after Capitalism* (San Francisco: Berrett-Koehler, 1999), 197.
4. "Social Justice in an Open World: The Role of the United Nations." https://www.un.org/esa/socdev/documents/ifsd/SocialJustice.pdf.
5. Wellbeing Economy Governments. https://www.gov.scot/groups/wellbeing-economy-governments-wego/.
6. "How Are Canadians Really Doing?" The 2016 CIW National Report. https://www.niagaraknowledgeexchange.com/wp-content/uploads/sites/2/2016/12/ciw2016-howarecanadiansreallydoing-1994-2014-22nov2016.pdf.
7. Adair Turner, "Monetary and Financial Stability: Lessons from the Crisis and from Classic Economics Texts." https://stage2.ineteconomics.org/uploads/papers/Turner.pdf.

Transform your life, transform our world. Changemakers Books publishes books for people who seek to become positive, powerful agents of change. These books inform, inspire, and provide practical wisdom and skills to empower us to write the next chapter of humanity's future.

www.changemakers-books.com

Current Bestsellers from Changemakers Books

Resetting our Future: Am I Too Old to Save the Planet?
A Boomer's Guide to Climate Action
Lawrence MacDonald
Why American boomers are uniquely responsible for the climate crisis — and what to do about it.

Resetting our Future: Feeding Each Other
Shaping Change in Food Systems through Relationship
Michelle Auerbach and Nicole Civita
Our collective survival depends on making food systems more relational; this guidebook for shaping change in food systems offers a way to find both security and pleasure in a more connected, well-nourished life.

Resetting Our Future: Zero Waste Living, The 80/20 Way
The Busy Person's Guide to a Lighter Footprint
Stephanie J. Miller
Empowering the busy individual to do the easy things that have a real impact on the climate and waste crises.

The Way of the Rabbit
Mark Hawthorne
An immersion in the world of rabbits: their habitats, evolution and biology; their role in legend, literature, and popular culture; and their significance as household companions.